AIRPLANES

Recent Titles in
Greenwood Technographies

AIRPLANES

THE LIFE STORY
OF A TECHNOLOGY

Jeremy R. Kinney
In association with the Smithsonian
National Air and Space Museum, Washington, DC

GREENWOOD TECHNOGRAPHIES SERIES

GREENWOOD PRESS
Westport, Connecticut • London

Library of Congress Cataloging-in-Publication Data

Kinney, Jeremy R.
   Airplanes : the life story of a technology / Jeremy R. Kinney.
      p. cm.—(Greenwood technographies, ISSN 1549–7321)
   Includes bibliographical references and index.
   ISBN 0–313–33150–2 (alk. paper)
   1. Aeronautics—History.   2. Airplanes—History.   I. Title.   II. Series.
   TL515.K45   2006
   629.1309—dc22   2006009759

British Library Cataloguing in Publication Data is available.

Library of Congress Catalog Card Number: 2006009759
ISBN: 0–313–33150–2
ISSN: 1549–7321

First published in 2006

Greenwood Press, 88 Post Road West, Westport, CT 06881
An imprint of Greenwood Publishing Group, Inc.
www.greenwood.com

Printed in the United States of America

The paper used in this book complies with the
Permanent Paper Standard issued by the National
Information Standards Organization (Z39.48–1984).

10 9 8 7 6 5 4 3 2 1

Contents

Series Foreword

In today's world, technology plays an integral role in the daily life of people of all ages. It affects where we live, how we work, how we interact with each other, and what we aspire to accomplish. To help students and the general public better understand how technology and society interact, Greenwood has developed *Greenwood Technographies*, a series of short, accessible books that trace the histories of these technologies while documenting *how* these technologies have become so vital to our lives.

Each volume of the *Greenwood Technographies* series tells the biography or "life story" of a particularly important technology. Each life story traces the technology from its "ancestors" (or antecedent technologies), through its early years (either its invention or development) and rise to prominence, to its final decline. obsolescence, or ubiquity. Just as a good biography combines an analysis of an individual's personal life with a description of the subject's impact on the broader world, each volume in the *Greenwood Technographies* series combines a discussion of technical developments with a description of the technology's effect on the broader fabric of society and culture—and vice versa. The technologies covered in the series run the gamut from those that have been around for centuries—firearms and the printed book, for example—to recent inventions that have rapidly taken over the modern world, such as electronics and the computer.

While the emphasis is on a factual discussion of the development of the technology, these books are also fun to read. The history of technology is full of fascinating tales that both entertain and illuminate. The authors—all experts in their fields—make the life story of technology come alive, while also providing readers with a profound understanding of the relationship of science, technology, and society.

Preface

This book is an outgrowth of lectures and ideas presented while serving as the Centennial of Flight lecturer at the University of Maryland at College Park in 2003. That would not have been possible without the financial, organizational, and institutional support of the departments of aerospace engineering and history at the university and the aeronautics division of the National Air and Space Museum. I would like to thank William Fourney, Robert Friedel, and Peter L. Jakab especially for enabling that wonderful experience. Additional concepts came from presentations for the MTU Aero Engines Summer Lecture Series in Paris, France, and the Aerospace and Mechanical Engineering Seminar at the University of Virginia, Charlottesville. National Air and Space Museum staff members also assisted me in the completion of the volume. Herbert Rochen, Dorothy Cochrane, Christopher Moore, Douglas Dammann, Kristine Kaske, and Kate Igoe provided research assistance, read drafts of the volume, and facilitated the selection and collection of the photographs. Herbert Friedman provided additional information.

Special thanks must also go to John D. Anderson, Jr., one of the deans of aerospace history, for his help, guidance, encouragement, and friendship over the years as well as in the completion of this project. As always, this book would not be possible without the graduate training I received at Auburn University from Stephen L. McFarland, James R. Hansen, and

William F. Trimble. Overall, I owe an incredible debt to the historians listed in the bibliography who have approached the history of the airplane in a critical and sophisticated way. The deficiencies in this book are solely mine.

Finally, I dedicate this book to Bonnie, whose encouragement and support of "the book" kept me motivated and made me think about the airplane in a broad perspective.

Introduction: Higher, Faster, and Farther

On the morning of Thursday, December 17, 1903, at approximately 10:35 a.m., the first successful airplane took to the air at Kill Devil Hill on the Outer Banks of North Carolina. At the controls was Orville Wright, who with his older brother, Wilbur, brought their Flyer from Dayton, Ohio, to the desolate windswept dunes as they had brought their gliders during the previous winters. This new machine was made up of two cloth-covered spruce wings with a horizontal elevator mounted in front and a vertical tail in the rear and an aluminum reciprocating piston engine and two wood propellers. The Flyer flew four times that day with Orville and Wilbur alternating piloting duties. Orville's first flight was for 12 seconds and traveled 120 feet. Wilbur's last flight around noon covered 852 feet for a duration of 59 seconds at 30 miles per hour. Before they could make another attempt, a strong gust of wind rolled the Flyer end over end, effectively demolishing it. Despite the loss, Wilbur and Orville knew full well that they were the first to design and construct an airplane capable of practical, sustained flight. They triumphantly sent their father a telegram:

> Success four flights Thursday morning all against twenty-one mile wind started from level with engine power alone average speed through air thirty-one miles longest 57 [sic] seconds inform press home Christmas

The first airplane: the Wright Flyer at Kitty Hawk, December 1903. National Air and Space Museum, Smithsonian Institution (SI 2002-16646).

Over the course of one cold and windy morning, the dream of powered flight had been realized.

Out of a modest beginning on those lonely Atlantic sand dunes, the airplane became a significant twentieth-century technology and continues as a major component of world society in the early twenty-first century. Along with automobiles, radio and television, electricity, computers, and medical advances such as anesthesiology, blood transfusion, and antibiotics, the airplane excited and amazed the world with the promise of progress through technology. The airplane eroded geographical boundaries such as oceans and continents, shortened the millenniums-old concepts of time and space, and fired the creativity and imagination of generations of individuals inspired to express themselves in both technical and cultural endeavors. On a higher level, the airplane enabled humankind to enter an environment formerly inhabited only by birds and beyond the atmosphere into space. With the airplane, humans went higher, farther, and faster than ever before.

The airplane also shocked and horrified the world when nations and political groups used it for military and political purposes. Whether it was a wood and wire contraption dropping small bombs on rebellious tribes in colonial empires, a single high-speed bomber dropping the first atomic bomb in history, the unseen stealth fighter sending smart bombs into enemy bunkers at night, or an airliner being used as an unwilling instrument of terror, the image of the airplane as a weapon of war has been with society as long as the image of it as an instrument of progress and peace.

During the early years of flight, many, including the Wrights, believed that airplanes would make wars impossible, but that proved to be a vain hope.

Both the positive and negative aspects of the airplane remind us of an old adage in the history of technology expressed by one of its founders, Melvin Kranzberg: "Technology is neither good nor bad; nor is it neutral." Essentially, a specific technology such as the airplane is not a living being capable of making history; it is an instrument of our own making and doing. Individuals, institutions, nations, and communities expressed their hopes, dreams, and ambitions through the airplane in its many forms. Keep in mind as you read this short book, that the airplane is a remarkable technology that stirs emotion, enthusiasm, and ambivalence. To understand the airplane is in many ways an attempt to understand how and why world society has embraced technology and used it over the past one hundred years. The airplane's use as an instrument of commerce, as a weapon of war, and as a vehicle for recreation reflects its adaptation to the modern world by humankind.

The story of the airplane is the story of the communities that created it for economic, political, military, and technological reasons, which reflects a mainstream cultural enthusiasm for technology overall. Airplanes are part of our everyday lives. They take us on vacation, on business trips, move precious cargo, and are instruments of peace and war, primarily redefining the way humankind travels, conducts commerce, spends its leisure time, and wages war.

From the very first flying machine, the propeller-driven wood-and-fabric Wright Flyer, to the latest military and commercial jets constructed from composite materials, all airplanes have one very important characteristic—they all are synergistic technologies that embody four primary systems: aerodynamics, propulsion, structures, and control. At the core of the airplane's success over the course of the twentieth century has been the development of these internal systems into an overall practical and symbiotic system. Without an equal balance between them, an airplane is incapable of flight.

Aerodynamics is a branch of fluid dynamics that deals with the movement of the airplane and the forces acting upon it in relation to the flow of air around it. The primary goal of aerodynamics is the creation of structures called airfoils, which generate lift, the upward force that allows an airplane to fly. An airplane wing is a series of airfoils along its entire length. The tail, or empennage, consists of a vertical and a horizontal stabilizer, which provide stability in flight. An equally important goal for the designer is the reduction of drag, the force that resists the forward movement of an airplane in the airstream. The design of an aerodynamically efficient airplane involves the use of sophisticated research tools such

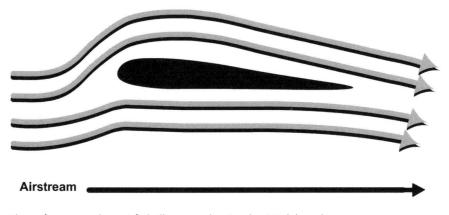

Airstream

Flow of air around an airfoil. Illustration by Sandy Windelspecht.

as wind tunnels and complex mathematical equations, which led observers early on to call aerodynamics the "science" of flight.

The primary responsibility of an airplane's propulsion system is to create thrust, the force that propels an airplane through the air. There are two main types of propulsion systems. The first consists of a propeller, an assembly of rotating wings, or blades, which converts the energy supplied by a power source into thrust to propel an airplane forward. The combination of a propeller and an internal combustion piston engine, similar to the ones found in everyday automobiles, was the main form of aerial propulsion for the first fifty years of flight. For speeds up to 500 miles per hour, the propeller and its power source are the most efficient because they can move a large mass of air at a low velocity, meaning they generate less waste. The second type of propulsion system, the gas turbine, or jet engine, is more efficient at speeds over 500 miles per hour. A jet engine takes in air, compresses it, mixes it with vaporized fuel, and ignites it to create thrust. After their revolutionary introduction during World War II, jet engines have become the dominant military and commercial source for propulsion.

The structure of an airplane, called an airframe, must be strong, and at the same time light in weight. The first airplanes were wood structures braced with wire for strength and covered in fabric for protection and aerodynamic airflow. Aircraft designers introduced metal and composite structures as well as different types of construction techniques over the course of the twentieth century as the performance of aircraft increased. Propeller-driven, piston-engine and jet aircraft have similar structural characteristics. The fuselage is the central structure that supports the lift-producing wings and tail and houses the pilots, passengers, cargo, instruments, engine, fuel, and landing gear. The wings and tail primarily house

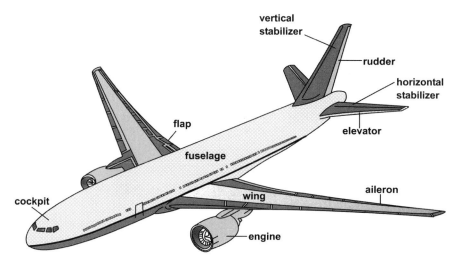

Parts of a jet airplane. Illustration by Sandy Windelspecht.

the control surfaces, but can also incorporate engines, fuel, and landing gear within their structures.

There is an important interrelationship between aerodynamics, propulsion, and structures that is crucial to the success of any airplane. Aircraft designers must maintain a balance between lift, drag, thrust, and weight, called the four forces of flight. For an airplane to take the air, the wings and engine must generate enough lift and thrust to overcome the weight and drag of its structure.

Control is the ability of the pilot to maneuver the airplane in the air by banking, rolling, climbing, and diving. Specialized structures control the

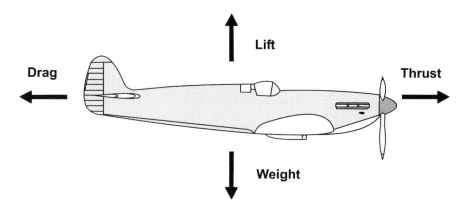

Four forces of flight. Illustration by Sandy Windelspecht.

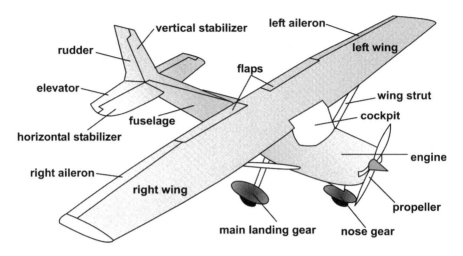

Parts of a propeller airplane. Illustration by Sandy Windelspecht.

airplane on three axes. The aileron, located on the outer trailing edges of the wings, provides lateral roll control and the ability to turn in flight. Mounted on the trailing edge of the horizontal stabilizer, an elevator provides longitudinal pitch control, or the ability to go up or down. The rudder, located at the rear of the vertical stabilizer, permits directional control from side to side, or in yaw. Flaps, mounted on the inside trailing edge of the wing, change the aerodynamic shape of the wing to create more lift during takeoff and landing.

This is a "technography" of the airplane that discusses its development from its origins in the late eighteenth century to the present day. It introduces its technological as well as its economic, political, and cultural importance in an international context with an emphasis on the United States. In the following pages, you will meet members of an international aeronautical community, the designers, engineers, entrepreneurs, military officers, pilots, and many others who created the airplane. Many of them will be famous individuals, but many more will be the unsung heroes that made flight possible. Enjoy meeting these people, but remember that they are members of a larger aeronautical family. The dramatic development and use of the airplane were the result of a communal response to challenges and concerns that tell us much about the societies that created them. Overall, the story of the airplane will be told over the backdrop of twentieth-century history.

The chapters in this volume survey eight major eras and themes in the history of the airplane. Chapters 1 through 3 discuss the introduction and early development of the airplane. Chapter 1 surveys the origins of powered flight, the creation of the first airplane, and its introduction as mainly a curiosity to the world during the period 1903–1914. World War I, discussed in Chapter 2, introduces the military airplane and an entirely new dimension of warfare called airpower. The Aeronautical Revolution of 1918–1938, presented in Chapter 3, resulted in the definition, creation, and refinement of the airplane into its modern form in the 1920s and 1930s. Those modern airplanes of the 1930s that fought World War II on a global battleground from 1939 to 1945 are discussed in Chapter 4.

The mid-twentieth century was a turning point in the history of the airplane. A new type of propulsion system, the jet engine, allowed airplanes to fly higher and faster than ever before. Chapter 5 details the reaction of the aeronautical community to that new technology, which resulted in a generation of new airplanes. This second aeronautical revolution was equal in importance to the original achievement of the Wright brothers.

After World War II, the airplane began to mature and take its place as an everyday technology. There would be significant aeronautical advances, but they improved upon a well-established system. Chapters 6 through 8 tell the story of the airplane from the perspectives of the three areas in which it has become an integral part. The arrival of the jet age coincided with the atomic age. In Chapter 6, military aviation rose to a new prominence as aeronautical technology became more sophisticated as the tensions of world politics increased. Commercial airliners, carrying people and cargo around the world, brought the world closer together and fostered a global industry detailed in Chapter 7. Chapter 8 documents the story of everyday, or general, aviation from its beginnings in the 1920s to the present. The volume concludes with a brief discussion of the airplane in the context of the first hundred years of flight.

This volume concludes with a glossary of technical terms and a selected bibliography. With the exception of reference files in the archives of the Smithsonian National Air and Space Museum, all materials are readily available for further research and inquiry by readers. This history is only a point of departure because there were so many stories and themes crucial to an understanding of flight that could not be included. Moreover, humankind has taken flight in types of craft other than the airplane. Helicopters, lighter-than-air balloons and dirigibles, engine-less gliders, and rockets and missiles are each worthy of their own technography.

Timeline

1000 BCE	Kites take to the air in China.
Around 1500	Leonardo da Vinci creates the first documented studies of flight and specific drawings of flying machines.
1799	Sir George Cayley creates his "silver disc."
1804	Cayley's 1804 glider.
1842	William Samuel Henson's Aerial Steam Carriage appears.
1853	Flight of Cayley's triplane glider with the family coach driver aboard.
1866	Formation of the Royal Aeronautical Society.
1870–1872	Francis Wenham creates the world's first wind tunnel.
1871	Flight of Alphonse Pénaud's Planophore.
1889	Otto Lilienthal presents his research in *Birdflight as the Basis of Aviation*.
1891	Samuel P. Langley's *Experiments in Aerodynamics* makes the first American contribution to aerodynamics.
1894	Octave Chanute publishes *Progress in Flying Machines*.

1896	Crash and death of Lilienthal. Chanute-Herring glider trials on the Indiana Dunes. Flight of Langley's Aerodromes No. 5 and 6.
1903	Second and last attempted launch of Samuel Langley's Great Aerodrome. The Wright Flyer makes the first powered, controlled, and heavier-than-air flight in history at Kitty Hawk.
1905	Flights of the improved Flyer, the world's first practical airplane, at Huffman Prairie near Dayton.
1906	Alberto Santos-Dumont makes the first powered, controlled, and heavier-than-air flight in Europe in the 14-bis.
1907	Organization of the Aerial Experiment Association.
1908	Henry Farman makes the first significant flights in a non-Wright airplane built by Gabriel and Charles Voisin. Glenn Curtiss takes to the air in his June Bug. The Wrights make demonstration flights in France and the United States.
1909	Louis Blériot crosses the English Channel in his model XI monoplane. The world's first air meet is held at Reims, France, west of Paris.
1911	Calbraith Perry Rodgers flies across the continental United States in the *Vin Fiz*.
1913	Marcel Prévost wins the Gordon Bennett Race. First aerial combat between two airplanes occurs during the Mexican Revolution.
1914	First airline begins regular scheduled passenger service.
1915	French pilot Adolphe Pégoud becomes the first ace. The Fokker Scourge begins due to the introduction of one of the world's first successful aerial weapons, the Fokker E series fighter.
1916	German pilot Oswald Boelcke formalizes the rules of air combat.
1917	The Germans conduct the first mass aerial bombing of a city when they attack London.
1918	Great Britain creates the first independent military air service, the Royal Air Force. Allied aerial offensives at St. Mihiel and the Meuse-Argonne.
1918–1925	The U.S. Air Mail Service provides the foundation for American commercial aviation.

1919	The U.S. Navy Curtiss NC-4 flying boat completes the first transatlantic crossing.
1920s–1930s	In the United States, the army, the navy, and the NACA work to change the shape, character, and use of the airplane.
1920s	Barnstormers roam the American countryside and entertain the public with the airplane. General aviation begins in the United States.
1920–1921	U.S. Army Air Service bombing tests prove the vulnerability of naval warships to aircraft.
1922	The first American aircraft carrier, USS *Langley*, enters service.
1924	The U.S. Army's Douglas World Cruisers circumnavigate the earth.
1926	Congress passes the Air Commerce and Air Corps acts to regulate and foster commercial and military aviation.
1927	Charles A. Lindbergh makes the first solo transatlantic flight. Jack Northrop's revolutionary Lockheed Vega takes to the air.
1929	Frank Hawks proves the value of the NACA cowling by flying the Texaco Lockheed Air Express from Los Angeles to New York. B. Melville Jones presents "The Streamline Airplane" to the Royal Aeronautical Society.
1930s	William J. Powell urges African Americans to pursue careers in aviation.
1932	Amelia Earhart becomes the first woman to fly the Atlantic solo. Jack Frye of TWA sends the letter that results in the Douglas DC airliners.
1934	The MacRobertson Air Race from England to Australia highlights advanced American aeronautical technology.
1935	The U.S. army air corps creates the GHQ Air Force to develop the tactics and technology required to wage strategic warfare. First flight of the Douglas DST/DC-3.
1937	German bombing of the Spanish town of Guernica. The Whittle Unit, or the W.U., becomes the first operating aeronautical gas turbine engine.
1939	Flight of the first jet airplane, the Heinkel He 178. German blitzkrieg into Poland.

1940	The Germans and the British fight the Battle of Britain.
1941	Japanese attack Pearl Harbor.
1942	Jimmy Doolittle leads the American bombing raid on Japan. The Battle of Coral Sea becomes the first naval engagement in history fought entirely by carrier-borne aircraft. First flight of the Messerschmitt Me 262. First flight of the Bell XP-59 Airacomet.
1942–1945	Allied Strategic Bombing Campaign in Europe.
1943	The WASPs go into service.
1944	Introduction of the North American P-51 Mustang in Europe.
1945	Boeing B-29s drop atomic bombs for the first time in history on the Japanese cities of Hiroshima and Nagasaki (August).
1945–1960	Propeller airliners pioneer international commercial aviation routes.
1947	Flight of the Bell XS-1 breaks the speed of sound.
1948	The first turboprop-powered airliner, the Vickers Viscount, takes to the air.
1948–1949	The Berlin Airlift.
1950–1953	U.S. North American F-86 Sabres and Soviet MiG-15s fight for air superiority over MiG Alley during the Korean War.
1952	de Havilland Comet I jet airliner enters service.
1952–1953	The area rule fuselage enables aircraft to fly regularly at supersonic speeds.
1953	Foundation of the Experimental Aircraft Association (EAA).
1955	Boeing B-52 Stratofortress enters service.
1958	Boeing 707 enters commercial service with Pan American Airways.
1959	Introduction of the Learjet, the first jet designed specifically for business aviation.
1959–1968	The North American X-15 program reaches the fringes of space.
1960	CIA pilot Francis Gary Powers is shot down and captured by the Soviet Union. Pratt & Whitney introduces the first practical turbofan engine, the JT3D.

1965–1973	American airpower fights the Vietnam War.
1969	The British military introduces the Hawker Siddeley Harrier.
1970s	Ascendance of the jet fighter-bomber in the United States.
1976	Concorde supersonic transport enters commercial service.
1977	Paul MacCready's Gossamer *Condor* becomes the first successful human-powered airplane with pilot Bryan Allen at the controls.
1978	Airline Deregulation Act comes into force.
1979	MacCready's Gossamer *Albatross* crosses the English Channel.
1981	Launch of the first space shuttle, *Columbia*, into space (April).
1986	Burt Rutan's Voyager makes the first nonstop circumnavigation of the earth without refueling.
1987	The Airbus Industrie A320 becomes the first civil airliner equipped with a fly-by-wire control system.
1989	The Lockheed F-117 Nighthawk enters combat during U.S. operations in Panama.
1990	The Lockheed SR-71 Blackbird, the world's fastest jet-powered airplane, flies across the United States in 1 hour and 4 minutes.
1991	Operation Desert Storm. Patty Wagstaff wins the first of three U.S. national aerobatic championships.
1995	The first "paperless" airplane, the Boeing 777, takes to the air.
1999	NATO airpower campaign stops the Serbian genocide in Kosovo, Yugoslavia.
2001	The Lockheed Martin F-35 Joint Strike Fighter STOVL prototype performs the famous "hat trick" at Edwards Air Force Base. Terrorists use hijacked jet airliners to attack the United States.
2004	NASA's X-43A hypersonic scramjet achieves Mach 10.

1

The Origins of Powered Flight, 1783–1914

◆

Two virtually unknown but extraordinarily gifted brothers from Dayton, Ohio, Wilbur and Orville Wright, invented the first flying and powered airplane in history, the wood, strut-and-wire-braced Flyer, and flew it in 1903. They synthesized over a century's worth of aeronautical knowledge and experimentation and combined that with their own innovative methods to create the first airplane. Their technology quickly spread across the Atlantic to Europe and merged with different approaches on how to design and build an airplane to create a new aerial age accompanied by a growing public fascination with flying machines. In September 1913, French air-racing pilot Marcel Prévost flew his Deperdussin monoplane to victory at the James Gordon Bennett Race at Reims, France, and became the first human to fly faster than 125 miles per hour. He flew a revolutionary new monoplane, the Deperdussin racer, with its molded plywood fuselage and powerful rotary engine. Just ten years before, Orville Wright made his first flight at a speed of 10 miles per hour at Kill Devil Hill. Those ten years witnessed a dramatic growth in the performance, the appearance, and the design process of the airplane.

THE PREHISTORY OF THE AIRPLANE,
1783–1896

The "dream" of flight is as old as human civilization. Various cultures from Asia, Africa, Europe, and the Americas included manifestations of human flight in folklore and legend. The most well-known expression of this idea comes from the Homeric oral tradition of ancient Greece and the eighth-century B.C.E. story of Daedalus and Icarus. Imprisoned on the island of Crete, Daedalus, a skilled craftsman, made wings of feathers and wax so that he and his son, Icarus, could escape across the sea like birds. The scheme worked, but Icarus, contrary to his father's warnings, flew too close to the sun. As a result, the wax melted and the ambitious young man fell to his death. The legend was a cautionary tale of humankind's aspirations to imitate the gods, but it was a persistent manifestation of the human urge to fly that remained in the Western imagination.

Soon human beings attempted to turn the inspiration of those stories into reality. Around the twelfth century A.D., a Benedictine monk in England, Eilmer of Malmesbury, constructed a pair of wings and launched himself from the local abbey at a height of 600 feet, breaking both legs in the process. The Malmesbury Cathedral commemorated his efforts as a tower jumper in a stained-glass window. Another English monk, Roger Bacon, described a birdlike, human-powered flying machine in his 1250 text, *Secrets of Art*. Constructing wings and attaching them to human arms and attempting to flap one's way into history was a persistent activity from the time of Eilmer until the late nineteenth century.

The first documented studies of flight and specific drawings of flying machines came from artist-engineer Leonardo da Vinci (1452–1519). The pages of his many notebooks illustrated an ornithopter, a human-powered flying machine with flapping wings based on the motion of a bird in flight, as well as the parachute and the helicopter. Nevertheless, human power and the direct replication of bird flight proved unsuccessful. At the time, however, there was no other conceivable means of solving the problem, which stymied experimentation for centuries. Da Vinci was the first to express belief that mechanical flight was possible and to implement a research program based on studying the problem through observation and rational thinking. Unfortunately, his fear of sharing his ideas led to the hiding and dispersal of his notebooks after his death until their discovery late in the nineteenth century.

While individuals dreamed of taking to the air like birds, there were seemingly unrelated, but more important, precedents for heavier-than-air flight due to their aerodynamic characteristics: kites, windmills, and toy helicopters. The first human-made objects to take flight, kites emerged in China around 1000 B.C.E. and quickly spread across Asia and the Pacific, and finally into Europe after the journeys of Marco Polo. By the fourteenth century, wind-driven mills, or windmills, served a variety of industrial duties to support the growth of modern Europe. The harnessing of the wind to generate power proved to be an inspiration for the toy helicopter, which appeared at approximately the same time. When spun by hand or by a string, the blades of the toy helicopter rotated at a speed fast enough to generate lift and fly upward rather than relying upon the wind to turn the blades like a windmill. Each of these technologies, seemingly mundane elements of everyday life, had a profound effect on the development of the airplane.

Another long-term movement in history contributed to the growing possibility of powered flight. The leading thinkers of the Scientific Revolution in Europe, uninterested in pursuing powered flight, nevertheless created its foundation, and the discipline of aerodynamics, through their creation of fundamental physical and mathematical principles during the seventeenth and eighteenth centuries. In England, the famous scientist Sir Isaac Newton (1642–1727) calculated and identified the forces of lift and drag through his work on the laws of motion. Swiss physicist Daniel Bernoulli (1700–1782) provided the basis for understanding the creation of lift when he revealed that air pressure decreased as the velocity of a fluid flow increased. Studies of windmills conducted by the famous English engineer John Smeaton (1724–1792) resulted in his identifying the advantage of cambered, or curved from front to back, surfaces over straight ones while traveling in the air. He used a new tool, a whirling arm, to test those surfaces. Smeaton's most celebrated contribution was his identification of the density of air for those tests in 1759. The figure became known as Smeaton's coefficient and would become a cornerstone of nineteenth-century aeronautical experimentation.

Despite the important antecedents to the airplane such as the mythic tale of Daedalus and Icarus, the drawings of Leonardo da Vinci, and the foundations of the Scientific Revolution, there was one individual who was directly responsible for what the airplane would become. Sir George Cayley (1773–1857), an English aristocrat and professional engineer from Yorkshire, originated the configuration of the airplane and highlighted the basic principles of flight. He was a man of many talents, especially in the newly emerging field of science and technology during the rapidly emerging Industrial Revolution. Cayley dabbled in agriculture, architecture, and railroad and watercraft design, and is considered the inventor of the caterpillar tractor, the bicycle wheel, and the prosthetic hand. As an activist for science and engineering, he helped create the British Association for the Advancement of Science and counted as his friends the leading scientists and industrialists of his day, including the railroad pioneers George and Robert Stephenson. Cayley was a product of a new age that required excellence in the ability to design and build new technologies and to theorize, analyze, and rationalize new ideas from a scientific viewpoint.

Cayley's central passion was aeronautics, and he began experiments with one of those small helicopter toys readily available in Europe in 1796. He believed that the ability of humankind to take to the air would be a tremendous contribution to world society. In 1799, he engraved upon two sides of a silver disc about the size of an American quarter two diagrams that reflected his ideas about flight. The diagrams would forever change how aeronautical enthusiasts would approach the creation of a practical flying machine. On one side, he drew an airplane design that separated the force of lift, generated by a wing, from thrust, which in Cayley's case was a woefully inadequate mechanical "flapper" propulsion system, and included a tail for control. The other side expressed the physical relationship between lift, drag, thrust, and weight for the first time in history. He combined those theories with practical experiments, which included a device that whirled wing airfoil shapes around and around so that he could calculate their efficiency. Cayley incorporated his concepts into his revolutionary hand-launched 1804 model glider, which, with its central fuselage, wings, and movable tail, became the first airplane of modern configuration to fly.

Cayley published his findings in three articles, titled "On Aerial Navigation," in the influential *Journal of Natural Philosophy, Chemistry and the Arts* from 1809 to 1813. Known as the "triple paper," the articles put into writing a solid framework for researchers to build upon later in the nineteenth century. Primarily, Cayley introduced five major principles. First, he recognized that a cambered wing produced more lift than a straight wing. The engineer identified the existence of air pressure on a wing—low pressure on the upper surface and higher pressure on the lower surface—that created lift on it. He presented the first analysis of the movement of the center of air pressure on wings in flight. Cayley recommended that moving the wingtips upward, above the centerline of the aircraft, created a dihedral angle to produce lateral stability. Overall, he demonstrated how to calculate aircraft performance. To demonstrate those concepts, Cayley built a successful unmanned full-size monoplane glider in 1809.

After 1813, Cayley turned to other areas of interest, but revived his work in aeronautics in 1849 by moving to full-scale experiments that produced the first successfully manned heavier-than-air flights. His tethered triplane glider took a ten-year-old boy into the air at his country estate. Four years later, his 1853 monoplane made a free flight with the family coach driver as the unwilling passenger. After a particularly hard landing, Cayley's driver remarked, "Please Sir George, I wish to give notice . . . I was hired to drive and not to fly."

Cayley is considered the founder of modern aviation. He was the first to use scientific and technological knowledge to address the problem of flight, especially the basic principles of aerodynamics. He theorized, designed, and built aircraft that would heavily influence the next generation of aeronautical experimenters.

Despite the hesitation of Cayley's coach driver, many individuals were anxious to take to the air in the nineteenth century, a period of intense international aeronautical enthusiasm in Europe. Quickly, two distinct approaches to building a practical flying machine arose. The first approach involved the immediate construction of full-scale, powered, and piloted aircraft based on rudimentary experiments. The second method viewed mechanical flight as a collective problem of lift, thrust, and control. These experimenters believed a crucial intermediate step to their success was full-scale gliding experiments by which they would gain valuable flying experience. Once that was mastered, then an adequate propulsion system would be added to an already successful flying machine.

In England, William Samuel Henson's design for a large passenger-carrying flying machine, the Aerial Steam Carriage (1842), took Cayley's original concept and added an engine and propellers. The prototypical airplane, if built, was to be as large as a modern-day jet airliner. The Aerial Steam Carriage was very popular with the public, and the wide distribution of prints featuring it illustrated symbolically what a flying machine should and would look like.

In France, Alphonse Pénaud, a brilliant young engineer with a world-class education, exhibited a fundamental understanding of what design elements aircraft would have, especially those related to longitudinal stability and balance. The 11-second flight of his small, rubber-powered airplane, the Planophore, at the Tuileries Gardens in Paris in August 1871, demonstrated that ability. In 1876, Pénaud designed and built a flying machine that resembled future aircraft with mechanic Paul Gauchot. The two-seat monoplane featured elliptical wings with cambered airfoils and a noticeable dihedral. The control system consisted of a control column connected to elevators and a rudder. The propulsion system was two tractor propellers turning in opposite directions. Perhaps the most futuristic aspect of the

Pénaud–Gauchot flying machine was its retractable landing gear. Unfortunately, circumstances prevented Pénaud from ever building the bold new design. Disheartened, ill, and facing ridicule for his efforts—it was a common belief that powered flight was impossible and that any individual attempting it was insane—Pénaud committed suicide in 1880 at the age of thirty. Nevertheless, he ranks with Cayley as one of the most influential aeronautical experimenters of the nineteenth century.

Clément Ader (1841–1926), a French engineer with a background in railroads and communications, designed and built the batlike Eole (1890), reflecting his study of flying animals. The tail-less airplane lifted off the ground under the power of a 20-horsepower steam engine and traveled 165 feet in October 1890, making it the first powered aircraft to take off from level ground with a human aboard even though it was incapable of controlled or sustained flight. The French Ministry of War gave Ader a substantial grant to construct a series of follow-on designs. After the failure of the Avion III to take to the air in October 1897, the French government withdrew its funding.

In England, American-born Hiram Maxim (1840–1916) was responsible for many new inventions, including the machine gun adopted by the European imperialist powers to control their territories in the late nineteenth century and to virtually annihilate each other in World War I. Not really interested in building a practical flying machine, he was interested simply in providing aerodynamic test data when he began experimentation with a massive whirling arm powered by steam in 1887. The equally huge biplane he unveiled in 1893 weighed 8,000 pounds and had a three-man crew. The gargantuan flying machine could never fly because Maxim's circular test track included an upper barrier that prevented it from rising higher than 2 feet. Within that barrier, the biplane traveled a distance of 40 feet at a speed of 42 miles per hour powered by its 180-horsepower steam engine. Maxim lost interest in further effort after the third test of his flying machine and stopped experimentation.

While these individuals embarked upon personal programs to solve the problem of flight, the first steps toward the institutionalization of flight were being taken. English scientists and engineers with unquestionable credentials formed the Royal Aeronautical Society in 1866, which was modeled upon the more-established scientific and engineering societies. The new group organized lectures, technical meetings, and public exhibitions, and published the influential *Annual Report of the Aeronautical Society* (1867) to spread the idea of powered flight. The creation of the Royal Aeronautical Society indicated that the idea of attempting to solve the problem of flight was practical.

One of the more influential members of the society was Francis Herbert Wenham (1824–1908). A professional engineer with a variety of interests, Wenham designed a glider with long, narrow (high-aspect ratio) wings that were similar to venetian blinds. Wenham is better known for creating the world's first operating wind tunnel in the period 1870–1872, which was the result of a grant from the Royal Aeronautical Society.

Germany's first aeronautical pioneer, Otto Lilienthal (1848–1896), took a different approach. He was a trained mechanical engineer and the only nineteenth-century flight experimenter with a college degree. In 1866, Lilienthal began aerodynamic testing and presented his data in *Der Vogelflug als Grundlage der Fliegerkunst*, or *Birdflight as the Basis of Aviation* (1889), which proved the superiority of cambered airfoils over flat plates and the importance of using mathematical coefficients to express aerodynamic data.

Lilienthal applied his research in aerodynamics to become the inventor, builder, and flier of the first practical hang gliders in history. He routinely glided through the skies in Germany and made over 2,000 successful flights from his man-made hill near Berlin. Lilienthal controlled his glider by shifting his weight while hanging from the glider in midair. He practiced the "birdman" approach to flight whereby he would learn to fly before attempting to add a power source. Lilienthal mistakenly believed that the key to powered flight was the construction of an ornithopter, which directly replicated bird flight with mechanical flapping wings. Before he could build his flying machine, however, he crashed one of his gliders in August 1896 and did not recover from his injuries.

The existence of the Royal Aeronautical Society and trained professionals such as Lilienthal added a considerable degree of cachet to aeronautical experimentation. These individuals reported their activities and discussed them at various scientific meetings, which enabled the community to stay up-to-date with the latest developments. The slow and gradual accumulation of a broader understanding of lift, control, structures, and propulsion contributed to a growing community of aeronautical enthusiasts.

At the core of this rapidly growing international community was Octave Chanute (1832–1910). Chanute was a French-born American civil engineer who had a very successful career building railroads in the western United States. His interest in aeronautics began in the mid-1870s for purely professional reasons. Chanute witnessed the influence of aerodynamics on the structural integrity of bridge spans and building roofs in high winds as well as the forward speed of locomotives. That intellectual curiosity soon turned into an intense obsession that turned him into one of the most important members of the new flying community. During the summer and fall of 1896, on the Indiana dunes off Lake Michigan, Chanute and his collaborator, Augustus M. Herring, flew gliders of their own design, including one reflecting the engineer's long experience with railroad bridge building. The "two-surface machine," with biplane wings and an overall structure resembling a bridge truss, was the first of the wood, strut-and-wire-braced aircraft that would dominate aviation in the first thirty years of flight.

Despite his enormous contribution to the development of aircraft structures, Chanute's greatest contribution was his role as a conduit between the major aeronautical experimenters of the late nineteenth and early twentieth centuries. He recognized the importance of the dissemination of knowledge and cooperation in solving technological problems. His book, *Progress in Flying Machines* (1894), was a synopsis of the work of the aeronautical community to that point and is considered the first work in aviation history. Chanute was a frequent presenter on flight to various groups. A presentation by a highly respected engineer—Chanute served as an officer in the American Society of Civil Engineers, the American Association for the Advancement of Science, and the Western Society of Engineers—was an important mark of credibility to aeronautical experimentation. Not only did he make individual presentations, but he also arranged for fellow aeronautical experimenters to go before the established engineering societies to inform the broader aeronautical community. Many of the well-known experimenters, including the Wright brothers and Samuel P. Langley, were helped immensely by Chanute's encouragement, sponsorship, and lengthy correspondence.

It was a Chanute presentation that inspired another early flight experimenter to become intensely interested in aeronautics. Samuel P. Langley (1834–1906), secretary of the Smithsonian and regarded as one of America's leading scientists, took on the challenge of solving the problem of heavier-than-air powered flight. The fact that Langley, a successful astronomer, scientist, and perhaps, above all else, administrator, became interested indicated the growing acceptance of aeronautical experimentation. Slowly, the introduction of extensive scientific testing and the reporting of data appeared in aeronautical experimentation.

Langley began his research into aeronautics in 1884 when serving as director of the Allegheny Observatory in Pittsburgh he continued his research after his appointment to the Smithsonian. He used a whirling arm to test flat plates, small model airplanes, and stuffed birds. His research culminated in 1891 with his book, *Experiments in Aerodynamics*, which stated emphatically that heavier-than-air powered flight was possible with existing technology. Langley's book was the first American contribution to the newly emerging field of aerodynamics.

Encouraged by his well-received work in theoretical aerodynamics, Langley started work on designing and building a practical flying machine. He started first by building models of the aircraft he intended to fly. His steam-powered model airplane, Aerodrome No. 5, launched from a catapult from the top of a houseboat in the Potomac River near Quantico, Virginia, in May 1896, became the first successful, unpiloted, heavier-than-air flying machine powered by an engine in history. The No. 5 featured a metal tubular fuselage with two sets of wood wings and a tail covered in silk. Within the fuselage, the 1-horsepower steam engine powered the two propellers, mounted between the two sets of wings, through a series of shafts and gears. The little aircraft weighed only twenty-four pounds. The No. 5 made two flights that day, the longest being 3,300 feet. Later, in November 1896, Langley's second flying model, Aerodrome No. 6, flew a longer distance of 4,790 feet.

Langley intended to just make a larger version of his model Aerodromes that would accommodate a pilot, namely, his assistant, the young engineer Charles M. Manly. He found support in the form of a $50,000 grant from the U.S. War Department, which was interested in new technologies during the Spanish-American War. Future U.S. president Theodore Roosevelt negotiated the deal when he served as assistant secretary of the navy. Langley's full-scale airplane, the Great Aerodrome, was ready in October 1903. Instead of a steam engine, the aircraft had a gasoline-powered radial engine modified by Manly to produce an astounding 53 horsepower. The first launching witnessed the crash of the airplane into the Potomac. A newspaper reporter present remarked that the failed airplane simply fell into the water like "a handful of mortar." Blaming the catapult mechanism, Langley and Manly prepared the Great Aerodrome for a second launching on December 8. Once again, the flawed structure of the Great Aerodrome crumpled under the force of the catapult and Manly found himself in the icy river again. Even if it had gotten into the air, Langley and Manly did not provide for flight control, one of the crucial technical systems of the airplane.

Langley was the one person that average Americans and the U.S. government believed would achieve heavier-than-air, powered, and controlled flight. His successful experimentation program, his published research, and the flights of the Aerodromes all appeared to be successful steps in that process. The spectacular failures on the Potomac in 1903 were dramatic, to say the least. After the second attempted flight, the War Department withdrew support, stating that the achievement of human flight was not yet possible. As a result, Langley received the scrutiny and ridicule of Congress, which helped reinforce the belief that flight was not possible. America's greatest scientist stopped his research, retreated into seclusion, and died a broken and unhappy man in 1906.

THE WRIGHT BROTHERS AND THE INVENTION OF THE AIRPLANE, 1896–1903

Unknown to the American public, during the same time that Langley's Great Aerodrome went crashing into the Potomac, two seemingly ordinary brothers from Dayton, Ohio, were reaching the culmination of their own sophisticated aeronautical research and development program on the Outer Banks of North Carolina. Unlike Langley's attempts, their efforts would result in the first practical airplane, which meant they achieved the first heavier-than-air, powered, and controlled flight in history. During their technological journey, they created the discipline of aeronautical engineering to go along with it. The brothers were Wilbur (1867–1912) and Orville Wright (1871–1948), two bachelors who were the successful owners of a bicycle business in the small midwestern town.

The Wrights came from a close-knit family headed by their father, a Protestant bishop, and their mother, who was both well educated and possessed a natural mechanical capability. The brothers, only four years apart, were very close. When they were young boys, their father gave them a toy helicopter, which fascinated them for hours on end. As young men, they were business partners. Their first joint venture, a printing company that featured a press of their own design, was highly successful and led to their work in the rapidly rising and lucrative bicycle market. Like many of their fellow citizens in the industrial heartland of the United States, the brothers were self-educated, self-employed, and mechanically inclined. Searching for a challenge, Wilbur read of Lilienthal's death in the local newspaper in 1896, which rekindled the interest in aeronautics begun by their childhood helicopter toy. Exhibiting a rational thought process from the very beginning, the brothers began an extensive literature search, which led them to the Smithsonian Institution and, more importantly, Octave Chanute. Wilbur and Orville believed that, at the least, they would contribute to the success of later experimenters who would finally build a practical airplane.

The engineering and research program the Wright brothers embarked upon would result in the world's first airplane. Building upon more than 100 years of previous research generated by Cayley, Pénaud, and others, they quickly ventured into their own pioneering territory. In the process, they created the discipline and the tools of aeronautical engineering. The

Wilbur and Orville Wright (left to right). National Air and
Space Museum, Smithsonian Institution (SI 2002-16615).

Wrights were uniquely qualified for the endeavor because they made an
excellent team. Wilbur excelled at theoretical and intellectual thought,
while Orville was a talented administrator and a capable partner for developing their mutual ideas during intense aeronautical discussions. Also,
they were especially skillful at applying technology that was readily available to them in a new and original way. Using what is called the "mind's
eye," they envisioned the airplane as a synergistic technical system based
on the four forces of flight—aerodynamics, control, propulsion, and
structures—and built a research program around that foundation.

The program initiated by Wilbur and Orville reflected their "birdman"
approach to solving the problem of heavier-than-air powered flight where
they would learn to design and fly a glider before adding a propulsion

system. They began rather modestly by building and flying a biplane kite in 1899. Unlike other kites that simply stayed aloft on currents of air, theirs incorporated a system of control based on their original idea of wing-warping that allowed them to maneuver their kite in the air. While talking to a customer in the bicycle shop, Wilbur flexed a long, rectangular paper box for a tire inner tube. As he talked, he twisted one end up and the other down, which allowed him to visualize a method of lateral control for their experimental aircraft. The wing-warping system would be the foundation of their control system on all of their subsequent aircraft.

The 1899 kite also exhibited another signature Wright technical characteristic. To an average engineer, the structure of the kite, and all later Wright aircraft with their struts and wire-bracing, resembled the type of construction found on railroad bridges across the United States called a Pratt truss. The use of the Pratt truss reflected the influence of Chanute, who used it in his earlier designs. The Wrights diverged from Chanute's totally rigid structure by removing bracing at the rear of the wings so that they could flex in the wing-warping system.

Achieving good results with the 1899 kite, Wilbur and Orville constructed a full-size glider the following year. The size of the aircraft, their desire to avoid public demonstrations of their experiments, and the lack of steady and consistent winds at speeds approaching 18 miles per hour in the Dayton area led them to the Kill Devil Hills near Kitty Hawk on the Outer Banks of North Carolina. The U.S. Weather Bureau advised them that it was the one place in the country that provided consistent, strong, and steady winds.

Despite the much-needed seclusion, both the 1900 and subsequent 1901 gliders performed poorly on the sandy dunes of the Outer Banks. Their wings did not generate enough lift and the gliders were uncontrollable overall. Frustrated, Wilbur and Orville began to reevaluate their aerodynamic calculations, which resulted in their discovery that Smeaton's coefficient, one of the early contributions to aeronautics, as well as Lilienthal's groundbreaking airfoil data, was wrong. They found the discrepancy through the use of their wind tunnel, a six-foot-long box with a fan at one end to generate air that would flow over small metal models of airfoils mounted on balances, which they had created in their bicycle shop. While Francis Wenham invented the wind tunnel, the Wrights were the first to use one in a systematic way that would result in a real flying machine and in the way that later aeronautical engineers would use it. The lift and drag data they compiled in their notebooks would be the key to the rest of their experimental program.

Wilbur and Orville used the new aerodynamic data to build their 1902 glider, which turned out to be the first successful integration of aerodynamics, structures, and flight control in history. Its biplane wings generated the lift the glider needed to stay in the air, and the Pratt truss structure held it all together. The brothers solved the problem of lateral control by linking the vertical tail rudder to the wing-warping system, which made the aircraft easier to control from the pilot's hip cradle. The pilot controlled the pitch with the elevator mounted at the front of the aircraft operated by a lever in his left hand. All in all, the Wright control system enabled them to master the three axes of flight: pitch, roll, and yaw. The flying program at Kitty Hawk was successful and very exciting as the brothers took turns soaring, turning, and banking over the dunes. Wilbur and Orville recognized the significance of that important intermediate step and patented their 1902 glider, which emphasized their intention to master the art of flying before going on to build a powered flying machine.

The Wrights were ready to design and build a powered flying machine, or what they called a Flyer, for testing at Kitty Hawk. They used the knowledge gained from the 1902 glider program to create a much larger aircraft. The Flyer featured a wingspan of 40 feet 4 inches, a length of 21 feet, a height of 9 feet 4 inches, and weighed 750 pounds with the pilot aboard. They covered the wood wings, rudder, and elevator with muslin fabric and relied upon wood skids for landing gear.

The propulsion system was the one element left to master in the Wright aeronautical system. Still working as a team, they focused on the two major components of a propulsion system: the source of power, or the engine, and the transmitter of that power, the propeller. Orville and Wright Bicycle Company mechanic Charlie Taylor concentrated on the engine. A suitable lightweight and powerful engine was not commercially available, so they had to design their own. The Wrights calculated that their engine would need to generate 8 horsepower to achieve flight. Orville and Charlie designed and fabricated a more-than-adequate 12-horsepower, water-cooled, four-cylinder engine with an innovative and light cast aluminum crankcase.

Wilbur believed that knowledge related to the design of ship propellers would help them design the propeller for their powered flying machine. They quickly realized that no formal theories existed and that they would have to figure it out themselves. After many heated engineering discussions, Wilbur and Orville realized that a propeller was simply a rotating, twisted wing moving in a helical path. As a result, they used airfoil data calculated from their wind tunnel to design two propellers able to convert the energy of their 12-horsepower engine into thrust. Using a drawknife and hatchets, they shaped the two-blade propellers from two-ply spruce, covered them in linen, and sealed them with aluminum powder mixed in varnish. Overall, the two Wright propellers generated thrust at 66 percent efficiency, which Wilbur and Orville calculated would be enough to get their Flyer off the ground at Kitty Hawk. They connected the engine to the two propellers with a chain-and-sprocket transmission system.

With the technical system complete, they ventured off to Kitty Hawk in September 1903 for another season of experimental flying with one main goal: to achieve heavier-than-air, powered flight. They intended to launch their Flyer from a long rail they called the Grand Junction Railroad to ensure consistent takeoffs. After a series of delays, equipment break-downs, and a crash on December 14, Wilbur and Orville achieved their goal on December 17 with four successful flights. The airplane had been "born" and with it the fundamental ways in which engineers would design its descendants.

A NEW AERIAL AGE, 1903–1914

The Wrights ventured back home to Dayton and continued to improve their airplane system to make it a practical vehicle. In addition to the secluded location at Kitty Hawk, they developed a new flying field at Huffman Prairie, a large cow pasture east of Dayton, where anyone could witness the flights, including passengers on the nearby local rail line. They adapted the rail takeoff system to include a large weight and pulley to launch their new Flyers when winds were low.

Wilbur and Orville's 1905 Flyer was the first practical airplane in history. Slight modifications from the original 1903 airplane included a larger elevator and rudder spaced farther from the wings to provide better control and separate wing-warping and rudder controls. With this new airplane, the Wrights were achieving the first circles and flights of great length. During one flight in October 1905, Wilbur circled the field thirty times in 39 minutes for a total distance of 24½ miles. Believing they had quickly passed from the experimental to the practical stage, the Wrights turned their focus to receiving a patent for their invention, awarded in May 1906, and finding a commercial customer in North America or Europe. Fearing that someone would steal their ideas, they would not fly again for two and a half years.

While the Wrights stopped flying, others in Europe and North America started taking to the air in their own flying machines. The first flights in human history, in lighter-than-air balloons, had occurred in France in 1783. Skeptical of the reports documenting the Wright flights, the French aeronautical community thought it would be the first to put a powered, heavier-than-air flying machine into the air. Alberto Santos-Dumont (1873–1932), an expatriate Brazilian living in France and a well-known airship pioneer, achieved the first sustained and powered heavier-than-air flight in Europe in his 14-bis in October 1906. The airplane was a silk-covered pusher biplane with an elevator mounted in front of the wings, a wheeled undercarriage, and a wicker balloon basket for the cockpit. Unlike the Wrights' flights at Kitty Hawk and Huffman Prairie, thousands of people saw what resembled a large box kite during its brief forays into the air.

The French quickly made important contributions to early aeronautics. Henry Farman (1874–1958), an English-born former bicycle, motorcycle, and automobile racer, made the first significant flights in a non-Wright airplane built by Gabriel and Charles Voisin. Farman's airplane, similar in appearance to the Wrights' Flyer, did not have a wing-warping system. It featured ailerons, movable structures hinged on the trailing edges of the wing, which provided the same lateral control in a more efficient manner. Ailerons required a straightforward mounting and connection system to the controls that did not require making the entire wing flex. In January 1908, Farman flew a circular pattern of more than one kilometer in his improved airplane and won the 50,000 franc Deutsch-Archdeacon Prize. Ailerons

Henry Farman winning the 1908 Deutsch-Archdeacon prize. National Air and Space Museum, Smithsonian Institution (SI 89-19606).

had appeared on previous aircraft, but Farman was the first to use them on a successful airplane. Farman began one of the first formal flight training schools in 1909 and founded one of the first manufacturing companies, the Farman Aviation Works, in 1914.

For propulsion, the French pioneered the use of the rotary engine for aeronautical use. Brothers Louis and Laurent Séguin of the Société des Moteurs Gnôme, builders of automotive and industrial engines, set out in 1907 to design and build the first aeronautical rotary engine. Their engine, called Omega No. 1, ran in 1908 and flew in Henry Farman's airplane in April 1909. Conventional internal combustion engines turned a crankshaft connected to a propeller and relied upon water running internally through them for cooling. This type of engine rotated—propeller, cylinders, crankcase, and all—on a stationary crankshaft. The spinning of the entire engine facilitated engine cooling by air and removed the need for the heavy radiator, coolant, and related equipment. The shorter crankshaft, compact design, and air cooling system resulted in a much lighter engine, which generated higher power-to-weight ratios. The excellent performance of the rotary engine ensured its use on many record-setting and combat aircraft. Quickly, Farman achieved both endurance and distance records at the Reims aviation meet in August 1909.

While the rotary engine greatly increased the performance of aircraft, there were drawbacks to its design. First, it was expensive to manufacture. The high-quality steel used in its construction and the craftsmen talented enough to make it were not cheap. Second, the engines simply could not be made larger to generate more horsepower. Rotary engines were powerful and light in a certain power range, but as aircraft became larger and needed more horsepower, another type of engine would be needed. Finally, ordinary engine lubricants could not be used due to the rotary motion of the engine. Oil derived from castor beans, better known as a popular laxative treatment, was ideal for use on rotary engines. Pilots flying rotary engine-powered aircraft received those benefits.

In North America, Alexander Graham Bell, the inventor of the telephone, was an aeronautical enthusiast who conducted his own experiments in the 1890s and supported Langley's work. He organized the Aerial Experiment Association (AEA) in September 1907 with two college students, a young army officer, Lt. Thomas E. Selfridge (1882–1908), and motorcycle builder and racer Glenn H. Curtiss (1878–1930), who served as the group's director of experiments. The AEA produced three notable aircraft, which illustrated their attempt to emulate the designs of the Wrights with the addition of more powerful Curtiss engines. Selfridge designed the AEA's first powered airplane, Red Wing, named for the fabric covering of its biplane wings. It flew a short distance in March 1908 before crash-landing. The second aircraft, called White Wing, featured ailerons on the upper wing and made several flights in May 1908. Curtiss achieved international recognition at the controls of the next AEA design, the June Bug, when he won the Scientific American Trophy on July 4, 1908, for the first witnessed straight-line flight over one kilometer.

After the AEA disbanded in March 1909, Glenn Curtiss went on to be a major figure in early aeronautics. His Curtiss Manufacturing Company became the first major American aircraft manufacturer. Its most successful design was the Model D Headless Pusher, which was produced in large numbers beginning in 1912. The Wrights and the French heavily influenced the overall design of the pusher biplane with its Pratt truss structure and aileron controls, but Curtiss added his own refinements. The airplane featured a three-wheel landing gear system with the nose wheel fairly far forward of the pilot to prevent the aircraft from turning over. Curtiss devised his own control system with a steering wheel that he pushed and pulled for elevator control to go up and down and turned for rudder control to turn left or right. His shoulders moved a yoke to actuate ailerons for lateral control to roll the aircraft. Two foot pedals controlled the engine throttle and the wheel brake. The combination of the elevator with the vertical stabilizer at the rear of the aircraft resulted in the "headless" nickname. Combined with his water-cooled, inline "vee" engine (called that because the cylinder arrangement resembled the letter "V"), the Model D was a very successful and adaptable product. The addition of floats to the basic design in 1911 resulted in Curtiss inventing the world's first seaplane.

While those aviators were taking their first flights into the air, the Wrights began their own flight demonstrations in Europe and the United States. Many individuals believed that they could not fly and that their claims were unfounded. The Wrights refurbished their 1905 Flyer, which included placing the pilot and now a passenger upright on the wing. The flights quickly proved their capability. Wilbur made the first public demonstration of a Wright airplane over Le Mans, France, in August 1908. Over the course of six months, Wilbur sufficiently quelled any doubt over whether the Wrights had flown in 1903. Orville took to the air over Fort Meyer, Virginia, near Washington, DC, in September 1908 to demonstrate their airplane to the American military. Tragedy struck when a propeller failure resulted in a crash of the airplane, which seriously injured Orville and killed his passenger, Lt. Thomas E. Selfridge, the army observer. The veteran of the AEA became the first passenger fatality in an airplane accident. Orville returned a year later with the Military Flyer, which became the American military's first powered flying machine. In the process of these demonstrations, Wilbur and Orville became the first major celebrities of the twentieth century.

The emergence of the airplane influenced long-held ideas about time, space, and national borders. A leading British newspaper, the London *Daily Mail*, offered a prize for the first aerial crossing of the English Channel, long considered the natural border between the island nation and the rest of continental Europe. In a 37-minute flight on July 25, 1909, a wealthy French engineer, Louis Blériot (1872–1936), flew from Calais, France, to Dover, England, in his model XI airplane. His flight was the first crossing of a major body of water by an airplane, and it significantly jarred British attitudes about the island's traditional isolation from continental Europe.

Blériot's model XI was very different from the Wright airplanes and very similar to modern aircraft. He invested much of his fortune from manufacturing automobile headlamps to hire his own staff and develop this new aircraft. The model XI featured a single, or monoplane, wing with the propulsion system—a three-cylinder engine and a wood propeller—mounted at the front of the fuselage. Since the propeller "pulled" the airplane forward, it was called a "tractor" monoplane. The Blériot control system consisted of a stick to manipulate pitch and roll while rudder pedals actuated yaw. The interconnected system allowed perfect control of the three axes of flight and persists to this day. The model XI landing gear consisted of two large wheels mounted under the wings and one small wheel at the tail. It offered ease of movement on the ground and facilitated taking off and landing on a variety of surfaces. Like the Wrights, however, Blériot used a wing-warping control system instead of ailerons and wood strut-and-wire bracing and fabric covering for the structure of the aircraft.

The Wright demonstrations and the near-hysteria generated by the appearance of an airplane over populated areas led to the organization of large air shows, or meets, in Europe and the United States, where the leading aviators of the day competed for prizes for endurance, speed, altitude, and distance. The first meet was held at Reims, France, west of Paris, in August 1909. Aviator Hubert Latham won prize money for altitude and distance flying his Antoinette monoplane. Other shows sprang up in Paris and Berlin (1909), New York City and Los Angeles (1910), and Chicago (1911). The great air meets became a venue for aviators other than the Wrights and Curtiss to become celebrities. Daring aviator Lincoln Beachey (1887–1915) rose to fame by performing death-defying aerial stunts, including racing his Curtiss biplane against the famous driver Barney Oldfield and his race car. Harriett Quimby (1875–1912), one of the earliest female aviators and known for her purple flight suit, became the first woman to fly across the English Channel in 1912.

Wilbur and Orville Wright decided to expand their manufacturing business to include an exhibition team in 1910. The group of flyers would represent the company at the air meets and publicize Wright products while demonstrating the safety of aircraft. They established flying schools at Huffman Prairie and at Montgomery, Alabama, and trained both military and civilian pilots. One of those pilots, Arch Hoxsey, took former president Theodore Roosevelt for his first airplane ride in October 1910.

In 1911, one of the aviators trained by the Wrights, Calbraith Perry Rodgers, became the first person to fly across the continental United States. Flying a Wright EX airplane, named for his sponsor, Vin Fiz grape drink, Rodgers was competing for a $50,000 prize for the first transcontinental flight made within thirty days, offered by newspaper publisher William Randolph Hearst. Rodgers took off from the track at Sheepshead Bay Speedway in New York on September 17, 1911, and arrived at Long Beach, California, on December 10 after flying more than 4,000 miles in eighty-four days. During his unsuccessful attempt for the Hearst Prize, Rodgers crashed the *Vin Fiz* five times, had numerous mechanical difficulties, and broke his ankle on the next-to-last leg of his journey.

The spectacular air meets became the venue for a new specialized aeronautical event, air racing, which persists to the present day. Competing against Europe's top aviators, Glenn Curtiss won the Gordon Bennett Cup speed race, the first air race in history, in August 1909 in a biplane called the Reims Racer. By far, the most successful air racer was the 1912 Deperdussin design of Louis Béchereau (1880–1970). It was a tractor

The shape of future airplanes: the Deperdussin Monocoque Racer. National Air and Space Museum, Smithsonian Institution (SI 74-9251).

monoplane powered by a rotary engine like Blériot's XI model, but it featured a new and innovative fuselage construction method called mono-coque, or "one shell." Workers formed the fuselage from alternating layers of wood glued together, shaped them into a mold, and covered the entire assembly in fabric. The single wooden shell facilitated a smooth, stream-lined shape. The fuselage was very different from the strut–and–wire bracing pioneered by the Wrights and used by the rest of the aeronautical community. The skin of the fuselage itself bore the structural loads and was internally hollow. With daredevil pilots such as Marcel Prévost at the controls, the Deperdussin racers were unstoppable and an important pre-cedent for future aircraft.

2

The First War in the Air, 1914–1918

◆

During the summer of 1915, the German military introduced one of the first successful aerial weapons, the Fokker E series fighter. The single-seat airplane, called *eindecker* in German for its externally braced monoplane wood wing, featured a strong fabric-covered, tubular steel fuselage, and, most importantly, a single machine gun capable of firing through the propeller with an innovative synchronizer linkage. The first generation of German fighter pilots took these aircraft, developed new aerial combat tactics and techniques, and achieved unchallenged air superiority in the skies over Europe for more than a year. The British and the French air arms countered this deadly "Fokker Scourge" with a new generation of fighter aircraft and tactics that quickly regained for them an advantage during the fall of 1916. This balance of airpower shifted from one side to the other till the end of the war with advances in technology serving as the key to victory. From 1914 to 1918, the airplane grew from an entertaining curiosity into a full-fledged weapon of war that promised global repercussions for the rest of the twentieth century.

THE ORIGINS OF AIRPOWER

World War I, the first of two world conflicts in the twentieth century, raged from 1914 to 1918. France, Great Britain, Russia, and, later in 1917, the United States faced Germany and Austria-Hungary in what contemporaries called the "Great War." A major characteristic of this first truly modern war was trench warfare where large armies slaughtered each other with machine guns, mass artillery barrages, and poison gas during suicidal frontal assaults in futile attempts to break an entrenched stalemate. The airplane, along with other new technologies such as the armored tank, motortruck, poison gas, and the submarine, made modern war even more deadly and impersonal by expanding the range and destructiveness of combat operations to include civilians on a global scale. Airplanes performed an important, but limited, part in World War I.

Airpower, the use of military aircraft to achieve tactical, strategic, and even political goals, emerged during World War I. It consists of four interrelated activities for airplanes. Reconnaissance and observation aircraft gather data such as ground troop movements and locations needed to wage military campaigns. The ability to gather that information depends upon the achievement of air superiority, the control of the skies over a particular area, with fighter aircraft. Bombing aircraft attack soldiers on the battlefield, civilians in their homes, and the resources used by nations to wage modern war. Finally, through airlift and air mobility, transport aircraft deliver ground troops and critical war supplies and tanker aircraft allow aircraft to fly unlimited distances through aerial refueling. All of these functions of military aircraft, with the exception of the very last, appeared during the Great War. Airpower also includes the creation of an air defense network on the ground that attempts to prevent attacks, observation, and interdiction by enemy aircraft.

The world military had used airplanes before the outbreak of the war. The reconnaissance, aerial photography, and subsequent bombing of enemy Turkish positions during the 1911 Italian invasion of Libya in North Africa were the first air combat missions in history. During a 1912 French campaign to stop native rebels in Morocco, French aircraft located enemy encampments and kept its leadership in constant communication with field units. Two American mercenary pilots, employed both by the government and the rebel forces during the Mexican Revolution, were the first to face each other in aerial combat in 1913. The Balkan wars of 1912–1913 pitted the small air forces of Turkey and Bulgaria against each other in what would be a final proving ground for air war in Europe.

At the outbreak of the war in 1914, the warring nations had relatively small air forces. Germany and its ally, Austria-Hungary, had approximately 340 combat aircraft. France and Great Britain had 160 and 80 aircraft respectively. These first major air forces performed reconnaissance and observation duties for the massive land armies, which was the primary function of the airplane during World War I. The ability to identify enemy troop movements and direct artillery fire and land forces toward them inhibited movement by both armies and contributed to a stalemate where the battle line devolved into a series of mud- and vermin-infested trenches separated by a deadly "no-man's-land" across France from Switzerland to Belgium. Above the trenches on both sides was a long line of tethered observation balloons, which, with the airplane, became the "eyes and ears" of the armies. Quickly, it became clear that the key to breaking the stalemate was the control of the air. Over the course of the war, one side would have an advantage with both technology and tactics only to be countered by rapid adaptation from the other in a continual tipping of the balance of airpower.

The first military airplanes to fly over the trenches were early flight aircraft pressed into military service to observe and photograph troop movements. The main German observation airplane was the Taube, or Dove, designed by Igo Etrich (1879–1967) in 1909. The Dove carried two passengers, a pilot and an observer, and featured a distinctive externally braced wood monoplane wing designed to appear like a bird's wing. The British flew their prewar design, the B.E.2C tractor biplane, designed and constructed at the government's Royal Aircraft Factory. The French flew outdated aircraft, such as the Caudron G.4 with its wing-warping lateral control, two 80-horsepower Le Rhône rotary engines, and a light structure and limited visibility. These observation aircraft would pass each other and the crews would simply wave or salute each other. As the seriousness of the war set in, the crews of observation aircraft would trade shots at each other with pistols, rifles, and even shotguns as they passed each other in the sky. During the first months of the war, pilots began to mount machine guns on their aircraft. In October 1914, two French fliers achieved the first aerial victory against another airplane by shooting down a German Taube with a machine gun mounted to fire forward from their Voisin Type 3 pusher biplane with the engine and propeller pointing backward like the Wright Flyer.

It became apparent to both sides that they needed to deprive their enemies of the ability to observe and record their activities, which led to the creation of the fighter airplane, the idea of air superiority, and the "ace." Roland Garros (1888–1918), an early flight pilot well known for his skill as an athlete, added to his Morane–Saulnier Type L monoplane a machine gun and triangular metal wedges to the propeller in April 1915. The most effective way to fire a machine gun at another airplane in the air was to mount it in front of the cockpit so that the pilot would be able to "point" the airplane at the target. That was easy to do for a pusher airplane, but they suffered from poor performance and were structurally weak due to their open fuselage frames. For tractor aircraft with better performance and stronger enclosed fuselages, the whirling propeller was in the way. Garros's metal plates would deflect the bullets away from the propeller, but still give him the accuracy needed to destroy German aircraft. His destruction of German aircraft in aerial combat proved the potential of a new type of airplane designed specifically for destroying others: the fighter.

Along with the new type of airplane came a new type of aviator, the fighter pilot, and more importantly, the ace. Generally speaking, an ace was a military pilot with five or more aerial victories over enemy aircraft. The title referred to the top card from a deck of playing cards. French newspapers used the phrase, "l'as de nôtre aviation," where the "as" (ace in English) was a popular term meaning "top" or "best," to describe one of Garros's fellow Frenchmen, Adolphe Pégoud (1889–1915), the first pilot to bring down five aircraft in aerial combat. Pégoud, a famous early flight pilot known for his spectacular aerial demonstrations, including the first "loop-the-loop," used the deflector system to destroy six German aircraft before dying in combat.

Despite the crude construction of the French deflector system, the Germans were at a technical disadvantage in the new aerial battlefields, but that would be short-lived. Ground fire forced Garros down behind German territory just three weeks after installing the deflector plates. German troops quickly captured him before he was able to destroy his airplane. Garros's fighter arrived at the Fokker aircraft factory in Schwerin, Germany, with an official order stating that the small company incorporate the deflector plates into its new E series monoplane. Designed as a single-seat scout airplane like the Morane–Saulnier, the E series featured an innovative tubular steel fuselage that was strong and light, pioneered by one of the company employees, Reinhold Platz (1886–1966). The head of the company, Anthony Fokker (1890–1939), a Dutch citizen living in Germany, had been building aircraft since 1913. He ordered his workers to develop an even better system than the deflector plates. They came up with a mechanical interrupter system that synchronized the machine gun to turn off when the propeller blades passed in front of it. The new system ensured that the Germans would have an immense technical advantage over both French and British aircraft.

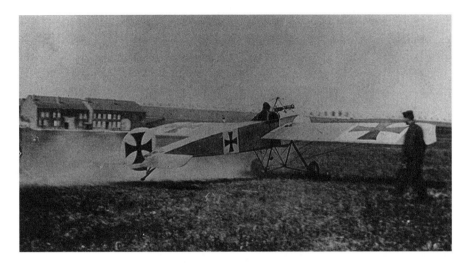

The world's first successful fighter airplane: the Fokker Eindecker. National Air and Space Museum, Smithsonian Institution (SI 90-10639).

The E fighter series was the first successful military airplane. The first Allied airplane fell to the single machine gun of the Eindecker in August 1915. What resulted was the infamous "Fokker Scourge" that raged until April 1916. German pilots flying the Fokker E fighter achieved complete air superiority. The Eindecker with its tubular steel fuselage and monoplane wing was superior in all respects to the aircraft flown by the British and French pilots, but it was the machine gun synchronizer apparatus that made the major difference. Allied pilots and aircraft were falling in large numbers to the guns of the Eindeckers. Soon, the British and French pilots were calling themselves "Fokker Fodder," meaning they were just there to be shot down by the Germans. They countered with the use of pusher aircraft that required no modification such as the de Havilland D.H.2, and better high-performance tractor biplanes, primarily the Nieuport 11 "Bébé" (Baby), which featured a machine gun mounted atop the wing to fire outside the path of the propeller. The pilot had to reach up to reload the gun during flight, which was a very difficult process.

The French and British, however, would soon have their own machine gun synchronizer systems. An Eindecker became lost in heavy fog and landed at the wrong airfield, where it was promptly captured. New aircraft, such as the French Nieuport 17, featured a machine gun synchronizer and a more powerful engine that proved to be more than a match for the rapidly obsolete Eindecker. In response, German pilots began flying a new generation of fighter aircraft, the Albatros D series biplane, which featured an elliptical wood veneer semi-monocoque fuselage with its high power, water-cooled, in-line Mercedes engine buried within the nose and a spinner covering the propeller to make an even, and streamline, contour throughout the airplane. The Albatros fighters used speed and firepower to outperform the lighter, but highly maneuverable, Allied aircraft.

While superior technology certainly gave an air service an advantage, it was the use and employment of the new fighter pilots that made the ultimate difference. A young German ace named Oswald Boelcke (1891–1916) formalized the basic rules of fighter combat in 1916 that are still taught to this day. Overall, he urged that fighter pilots use any and all advantages they could to destroy a target. Attacking pilots were to use tricks such as keeping the sun behind them so that the glare of the sun would disorient their opponents. That enabled them to dive upon their targets and fire their machine guns at point-blank range when it was obvious they would hit their target. Once engaged, fighter pilots were to follow through with an attack, keep the opponent within sight at all times, and never be deceived by clever enemy tricks. If attacked, fighter pilots were to fly to meet the enemy rather than attempting to evade the attack, and if over enemy lines, fighter pilots were to have a specific line of retreat. For groups of aircraft, Boelcke dictated that aircraft attack in groups of four or six with each taking on a specific opponent when the dogfight degenerated into a series of single pilot-to-pilot combats.

In 1916, Boelcke was Germany's leading ace with forty victories and he was a national hero. The government gave him its highest military award, the Pour le Mérite, better known by its nickname, the "Blue Max," and the opportunity to spread his ideas by forming a dedicated fighter squadron, or Jasta. Promising new pilots in the elite unit, who would also become aces and wear the Blue Max, included Werner Voss (1897–1917) and a young aristocrat named Manfred von Richthofen (1892–1918), a baron from Prussia in eastern Germany. Von Richthofen's natural skills as a hunter, his service in the cavalry before entering aviation, combined with his training from Boelcke, made him a masterful fighter ace who focused on one goal: killing the enemy.

Von Richthofen took command of his own Jasta after both Boelcke's and Voss's deaths in January 1917. To set himself apart from other pilots, he ordered that his entire Albatros fighter be painted a bright crimson, which made him instantly recognizable in the air and earned him the infamous nickname, the "Red Baron." He led the unit during "Bloody April" when a small number of superior German pilots and aircraft massacred a numerically superior British force flying outdated equipment. Von Richthofen went on to command a group of fighter squadrons (*Jagdgeschwader*) which, in the spirit of their leader, flew brightly painted aircraft, earning them the nickname "The Flying Circus." The Red Baron's prowess as a fighter pilot made him a household name in both Europe and North America, especially after the appearance of his 1918 autobiography, *Der rote Kampfflieger* (*The Red Air Fighter*). Von Richthofen achieved a total of eighty confirmed victories, the highest of any pilot during World War I, before his death in April 1918. During his last months, the Red Baron and his Flying Circus flew the Fokker Dr.I triplane, which became forever linked to him, but the majority of his kills were made in Albatros biplanes.

During that spring when the Red Baron fell for the last time, the British and the French had regained air superiority over Germany. Throughout the remainder of the war, the British, French, and the Germans fought the ever-changing war of tactics and technology. Aircraft designers in each nation had many choices to design a high-performance fighter, which resulted in a variety of approaches. The British and the French answered with conventional wood strut-and-wire-braced airplanes while the Germans answered with an increasing number of innovations.

The single-seat Sopwith Camel was the first British fighter with two machine guns synchronized to fire through the propeller when it entered combat operations in July 1917. The Camel's 130-horsepower Clerget engine represented the upper realm of rotary engine development and generated a modest top speed of 105 miles per hour. Developed from a very successful family of fighters starting with the Pup in 1916, the Camel featured a metal covering over the rear of the guns that reminded pilots of its namesake animal's "hump." The concentration of the heaviest components of the aircraft—the engine, fuel, guns, and pilot—into the front combined with the gyroscopic force created by the rotary engine and the resultant difficult manipulation of the flight controls made the Camel highly maneuverable, but very dangerous to fly for newly trained pilots. Despite the hazards, aviators flying Camel fighters destroyed almost 1,300 enemy aircraft during the war, more than any other Allied pilots.

The French used a highly successful series of fighters produced by SPAD (or *Société pour l'Aviation et ses Dérives*), which was a reorganization of the company that produced Louis Béchereau's Deperdussin monoplane racer. Béchereau's SPAD fighters were wood and fabric, strut-and-wire-braced biplanes with exceptionally thin airfoils, reflecting the norm for aircraft since the time of the Wrights. The final version, the SPAD XIII, entered service in May 1917 and was capable of a maximum speed of 130 miles per hour and carried two machine guns synchronized to fire through the propeller. All three of the highest-scoring French aces—René Fonck (75), Georges Guynemer (54), and Charles Nungesser (45)—and the majority of the French escadrille, or fighter squadrons, flew SPAD aircraft. The high speed, formidable armament, and group tactics used by the French made the SPAD XIII more than a match for any German aircraft.

The engine was the key to the high speed of the SPAD. Béchereau chose the most powerful aircraft engine of the time for his fighters, an in-line water-cooled eight-cylinder "vee" engine capable of generating more than 200 horsepower manufactured by the Hispano-Suiza company near Paris. The company was well known before the war for its high-quality automobiles and company engineer, Marc Birkigt (1878–1953), took one of his engine designs and adapted it for aeronautical use. Instead of individual steel cylinders bolted to a heavy, bulky crankcase, he developed a cast-aluminum cylinder row that incorporated four pistons, a neatly designed valve mechanism, and internal water cooling passages and attached them to an alloy crankcase. The Hispano-Suiza engine was compact, light in weight, and powerful, and it influenced an entire generation of engine design.

French ace Georges Guynemer at the controls of his SPAD fighter. National Air and Space Museum, Smithsonian Institution (SI 2003-10830).

The most successful German fighter of World War I was the Fokker D.VII, which was a significant departure overall from other aircraft of the period. Reinhold Platz, as chief designer, and Anthony Fokker, excelling in the role of company test pilot, worked together to produce an exceptional fighter that outclassed other German designs in a 1918 government competition and received the personal endorsement of von Richthofen, the Red Baron, himself. Like the Eindecker, the D.VII had a welded tubular steel fuselage, but it did not have thin, externally braced wings like its monoplane predecessor. The wings of the D.VII were internally braced, or cantilevered, meaning that no outside supports such as wires or struts were needed to keep them in place. Fokker and Platz incorporated the struts mounted between the upper and lower wings to alleviate any anxiety displayed on the part of the pilots on not having any "support." As a result, the D.VII used revolutionary new thick airfoils that were a significant departure from the thin airfoils used by the Wrights and their predecessors. Thick airfoils generated more lift, which offered a significant performance advantage for all types of aircraft. For fighter aircraft, thick airfoils enabled a tighter turning radius that made them more maneuverable. The combination of a tubular steel fuselage, cantilever wings with thick airfoils, and a water-cooled 160-horsepower six-cylinder Mercedes in-line engine gave the D.VII unprecedented performance and maneuverability and enabled it to reach a speed of 120 miles per hour at altitudes up to 21,000 feet.

The D.VII reached German combat units in April 1918. Allied pilots mistook the D.VII's plain appearance to be an indicator of poor performance, but quickly realized they were facing a deadly, and highly effective, enemy. German pilots could make the D.VII virtually "hang" on its propeller in midair—a combat maneuver that no other airplane was capable of performing—to fire its two machine guns at British and French aircraft in their most vulnerable positions: behind and below. At the conclusion of the war, the victorious Allies feared and respected the D.VII so much that the peace treaty signed at Versailles included a provision that the Germans turn over all of the biplane fighters. The combat record and overall performance of the D.VII has led many authorities to argue that it was the best fighter airplane of World War I.

Despite the exploits of those daring aces and the increasing performance of their fighters as they fought for air superiority, the Great War was not an aerial war. It was a ground war where hundreds of thousands of soldiers fought back and forth across "no man's land" as they faced machine guns, mass artillery barrages, and poison gas. While fighter aircraft preyed on observation and reconnaissance aircraft over the front, there was another technology that ground troops had to specifically be afraid of: the ground-attack, or close-support, airplane. At the same time that pilots and observers began to attack each other in the air, it became obvious that aircraft could attack targets on the ground as well. It began simply enough with observers dropping bombs over the sides of their airplanes to attack troop concentrations on the battlefield, but soon a specialized type of aircraft and tactics emerged.

The use of aircraft in ground attack emerged on a large scale in 1917. During the Battle of Cambrai in November, the initial British attack included 300 aircraft, mostly fighters, ordered to strafe, or fire their machine guns at targets on the ground, and bomb German troop positions. As the British advance slowed down, the Germans counterattacked with airplanes used specifically to fight British ground forces in conjunction with special ground troops. The Germans used ground-attack aircraft to great effect for the remainder of the war and enjoyed considerable success with their use. These low-flying ground-attack aircraft, flying in groups of four to six at altitudes of less than 100 feet, harassed Allied artillery and troops with bombs and machine gun fire to help clear the way for German infantry. When those ground attacks failed, the aircraft would bomb and strafe troop assembly points to stall their offensives.

Two important German ground-attack aircraft were the Halberstadt CL.IV and the Junkers J 1. The Halberstadter Flugzeug-Werke (Halberstadt Aircraft Factory) designed the plywood-skinned CL.IV biplane specifically for ground attack. The pilot and rear gunner shared one large cockpit that facilitated close communication during missions. The pilot had two fixed, forward-firing machine guns to use for strafing while the gunner had at his disposal a supply of antipersonnel grenades and small bombs to attack troops and a single machine gun on a flexible mounting for defense against enemy fighters. Powered by the same Mercedes engine used in the Fokker D.VII, the CL.IV was highly maneuverable and fast with a top speed of 112 miles per hour when it entered combat in April 1918. As Allied air defenses became more deadly, it became apparent that the CL.IV could not survive in such a hostile environment with its maneuverability and speed alone. It needed armor.

The need for protection from ground fire led to the inclusion of metal in aircraft construction. Dr. Hugo Junkers (1859–1935), an engineering professor and successful manufacturer of home-heating products, believed in the importance of the cantilever, or internally braced, wing, which led to his adoption of a thick wing to house the internal support structure. The aerodynamic and structural advantages complemented each other. His first airplane, the monoplane Junkers J 1, flew in December 1915. The wings, covered in corrugated and flat iron sheets, proved to be the first cantilever, stressed-skin wings capable of carrying an aerodynamic load. The J 1 exhibited sluggish performance, but the German government was interested in the "armored" airplane. For his next design, Junkers used a new construction material, duralumin, a strong and light aluminum alloy. The improved design, the J 4 biplane, the first all-metal airplane produced in quantity, followed quickly. The J 4 was slow and poor in performance, but the German government believed that its metal construction would make it an ideal observation and ground-attack airplane.

THE FUTURE OF AIRPOWER

Another critical component of airpower, strategic bombing, emerged during World War I. While observation pilots collected data and targeted artillery, fighter pilots fought for air superiority, and ground-attack pilots harassed targets on the battlefield, strategic bomber pilots attacked the ability of an enemy nation to wage war. Targets included resources such as factories and military installations, infrastructure such as government institutions, utility providers, and transportation centers, and, for the first time, civilian populations. In previous conflicts, civilians had suffered as a result of naval blockades and military sieges, but they had never been attacked directly as part of a strategic campaign. Military leaders hoped to erode the psychological well-being of a nation by bombing its citizens, which raised highly contentious ethical and moral issues that are still discussed to this day.

The idea of strategic bombing, especially the idea of attacking civilians, was untried, but it was not new. H. G. Wells forecasted the aerial destruction of New York City in his 1908 book, *The War in the Air*. In the opening months of the war, individual and small groups of aircraft on both sides had attacked civilian targets and there was small loss of life. From January 1915 to August 1918, the Germans initiated the first dedicated strategic bombing campaigns when they attacked England with lighter-than-air dirigibles, called zeppelins after the German count that pioneered their use in Europe. The Germans suffered heavy casualties and loss of equipment for the damage they inflicted on the British.

The first major bombing campaigns carried out with aircraft occurred over battlefields far from France. On the eastern front, the Imperial Russian government organized the Eskadra Vozdushnykh Korablei (Squadron of Flying Ships), an independent operational bombing unit of forty aircraft, in February 1915. Designed by pioneer Russian aircraft designer Igor I.

Sikorsky (1889–1972), the bombers, called Il'ya Muromets after the epic hero who saved Russia from a Mongolian invasion, were the world's first four-engine aircraft when they first flew in 1913. The Il'ya Muromets was a large strut-and-wire-braced biplane that carried a crew of five and was capable of carrying 1,543 pounds of bombs. The giant bombers conducted a successful bombing campaign against German forces, especially against critical rail junctions, until Russia's withdrawal from the war due to the Bolshevik Revolution.

On another front far from France, Italy entered the war against Germany and Austria-Hungary in May 1915. The Italian Corpo Aeronautica Militaire immediately began a strategic bombing campaign against Austro-Hungarian forces that continued until the end of the war. The Italians used massive twin- and tri-motor bombers with triplane wings designed by Count Giovanni Battista Caproni, who, like Sikorsky, had built his first large airplane in 1913. Both the British and the French used Caproni aircraft over the course of the war. New ideas about the use and employment of strategic aircraft emerged from an obscure Italian military officer, Giulio Douhet (1869–1930). Even though he never learned to fly, he became a pioneer airpower theorist through articles, a novel *How The Great War Ended—The Winged Victory* (1918), and his landmark treatise on air warfare, *Command of the Air* (1921).

Frustrated with the poor results with zeppelins, the Germans used multiengine bombing aircraft to attack British and French cities beginning in May 1917, with the first attack on London occurring a month later. More than 400 multiengine bombers—Gotha twin engine bombers and

An Il'ya Muromets of the Imperial Russian Eskadra Vozdushnykh Korablei. National Air and Space Museum, Smithsonian Institution (SI 83-16519).

Zeppelin Staaken four engine bombers—conducted more than fifty raids that inflicted almost 6,000 casualties at the cost of sixty-two aircraft. The British government reacted to the German air raids by combining the army's Royal Flying Corps and the navy's Royal Naval Air Service to create the Royal Air Force (RAF), the world's first independent air arm, in May 1918. Part of the new organization was a strategic component where RAF Handley-Page bombers attacked a variety of targets during the war, including railroad centers, factories, and submarine bases. Along with the RAF, the British created the Air Ministry, the first government agency dedicated to civilian, commercial, and military aeronautical matters. The chief of the RAF was Gen. Hugh "Boom" Trenchard (1873–1959), who intended, had the war persisted into 1919, to attack German cities with new four-engine Handley-Page bombers.

AMERICAN MILITARY AVIATION AND THE GREAT WAR

The United States was by far the least prepared for conducting an aerial war. The British, French, Germans, Italians, and Russians all had three years of war to stimulate aeronautical development. During the years leading up to World War I, the United States ranked thirteenth in governmental aeronautical appropriations. Both Brazil and Bulgaria spent more money on military aviation than the United States.

For the American military, the early flight era was a time for exploring the possibilities of flight. These flights would be indicators of the capabilities of future military aircraft. Pioneer aviator, Eugene Ely (1886–1911), made the first takeoff from a ship, the cruiser *Birmingham* anchored off Hampton Roads, Virginia, in his small Curtiss pusher biplane in November 1910. Later in January 1911, Ely landed and took off from the cruiser *Pennsylvania* at anchor in San Francisco Bay.

While Ely demonstrated the promise of naval aviation, the first generation of army pilots experimented with military applications of the airplane. In August 1910, Lt. Jacob Fickel, riding as a passenger with Glenn Curtiss near New York City, became the first person to fire a rifle while airborne. Lt. Myron Crissy dropped the first live bomb from an airplane in January 1911 near San Francisco. Later in June 1912, Col. Isaac Lewis, inventor and namesake of a new air-cooled machine gun, tested his new weapon over the army training field at College Park, Maryland.

Volatile political events south of the U.S. border led to the first American military use of the airplane. When President Woodrow Wilson dispatched the capital ships *Mississippi* and *Birmingham* to the Mexican port of Vera Cruz in April 1914, five Curtiss flying boats searched for underwater mines and scouted the nearby countryside. On the border, the bandit-revolutionary Pancho Villa staged a raid on Columbus, New Mexico, which resulted in the deaths of seventeen Americans. In retaliation, Brig. Gen. John J. Pershing led two columns of American troops into the high mountains of northern Mexico in pursuit of Villa from March 1916 to February 1917. Accompanying the troops was the First Aero Squadron under the command of Capt. Benjamin D. Foulois (1879–1967). Pershing used the squadron for reconnaissance and as a rapid courier service between his ground units, but logistical, organizational, and technical difficulties highlighted the shortcomings of American military aviation.

Individual Americans had been fighting in the air war with the French, British, and Italians before April 1917, which provided a limited legacy of experience for the air service. The French called their more than 200 American volunteers the Lafayette Flying Corps and operated one all-American squadron, the Lafayette Escadrille (named after the French general who helped the American colonists win their freedom from England during the American Revolution) which became world famous. The squadron's leading ace, Raoul Lufbery (1885–1918), achieved seventeen victories. After America entered the war, the Lafayette Flying Corps dissolved and its members joined newly arrived American units in France. One of the volunteer fighter pilots, however, stayed in foreign service. Due to cultural attitudes toward minority groups at the time, the air service would have never allowed African American Eugene Bullard (1894–1961), a proven fighter with the French, to fly in American units.

When the United States entered the Great War in April 1917, the army and navy had just over 100 aircraft and none of those were suitable for fighting an aerial war. Congress approved an ambitious aircraft production program in July 1917 backed by an unprecedented $640 million appropriation ($10 billion in modern currency) that would produce more than 100,000 aircraft by 1918. America was undergoing rapid industrialization that included the emergence of mass-production techniques characterized by Henry Ford's automobile assembly lines in Detroit, Michigan. Military planners simply believed that "good old American know-how" and "Yankee ingenuity" could be applied to aircraft production in the same manner to bring the United States up to an equal aeronautical level with Europe.

For the 1918 aerial campaigns, the U.S. government purchased European observation, fighter, and bomber aircraft for immediate use while it selected European designs for modification and production to go into service later. The United States intended to field combat aircraft of American design and manufacture for the 1919 campaigns. The dependence upon Europe was a glaring indicator of how far the United States was behind Europe in aeronautical technology.

The one European airplane that the United States settled upon and actually produced was the Liberty battle plane. It was the American version of the British DH-4 bomber and observation airplane designed by the highly successful engineer, Geoffrey de Havilland (1882–1965). After American entry into the war, patriotic citizens quickly applied the name "liberty" to any items associated with the war. Liberty bonds raised funds to help the government pay for the war; Liberty trucks carried troops and supplies to the front; and the Liberty battle plane was to clear the skies and level the battlefields in advance of victorious American ground troops. The military called it a "battle plane" because it was to serve in the multiple roles of artillery spotting, bombing, observation, photo reconnaissance, and, if needed, as a fighter. A British-built DH-4 and an American Liberty battle plane appeared to be the same aircraft—a large wood, strut-and-wire-braced biplane capable of 124 miles per hour with a liquid-cooled V-12 engine—but they were very different regarding the hundreds of internal changes the Americans made to the design to manufacture it in quantity.

America's contribution to the first war in the air: the Dayton Wright Liberty battle plane. National Air and Space Museum, Smithsonian Institution (SI 76-2204).

American manufacturers, primarily the Dayton Wright Company, of Dayton, Ohio, produced 4,500 DH-4s with approximately 700 reaching the battlefront in 1918. They were the only American-manufactured aircraft to serve with the army during the war. A standard American DH-4 was well armed with two fixed, forward-firing machine guns, two flexible machine guns for the observer, and six 25-pound bombs, as well as reconnaissance and communications equipment, including two cameras, and a radio transmitter. American pilots and observers called the DH-4 the "flaming coffin" due to the large fuel tank placed between their separate cockpits, which impaired communications and ensured that gasoline would be all over them in the event of a crash.

The Liberty battle plane had its own Liberty engine, which was the greatest aeronautical contribution the United States made to the war effort. The official designation was the U.S. Army Standardized Engine, but it was simply called the Liberty engine in the spirit of the time. The American government recruited two automobile engineers, Jesse G. Vincent (1880–1962) of the Packard Motor Car Company of Detroit, Michigan, and Elbert J. Hall (1882–1955) of the Hall-Scott Motor Company of Oakland, California, and made them engineering officers as part of its efforts to jump-start the American aviation production program. Vincent and Hall holed up in the Willard Hotel in downtown Washington, D.C., for one week in July 1917 and designed a family of liquid-cooled V-type engines in six, eight, and twelve-cylinder variants to power a planned new generation of American combat aircraft. The twelve-cylinder engine, rated at 400 horsepower, became the standard version because none of the planned aircraft, except for the Liberty battle plane, materialized. The major American automobile companies—Ford, Lincoln, Packard, Marmon, and Buick—produced 22,000 engines before the end of the war.

The American aviation industry could handle the production of homegrown training aircraft that would create a new generation of aviators. The Curtiss Aeroplane and Motor Company produced the JN series, known widely as the "Jenny." The original design was a combination of two aircraft designed by Glenn Curtiss and his new engineer, just hired from Sopwith in England, B. Douglas Thomas. The Jenny was a two-seat, wood, strut-and-wire-braced tractor biplane with a 90-horsepower Curtiss OX-5 V-8 engine. It had two sets of controls so that, if needed, a cautious instructor could take over control of the airplane from a less-than-proficient student. More than 90 percent of American aviators trained during the Great War made their first flights in the most successful version of the Jenny, the JN-4D trainer.

Despite the promise of American industrial capability, the United States exhibited a total reliance upon Europe for frontline combat aircraft. The American aviation industry was virtually nonexistent and it had no similarities whatsoever to the automobile industry that was capable of producing tens of thousands of cars. Automobiles and airplanes were very different technologies that facilitated different manufacturing techniques. The newly emergent American aviation industry, consisting of 300 plants employing more than 200,000 people produced only 12,000 aircraft, of which 8,500 were trainers, and 28,500 engines at a total cost of $848 million ($12 billion in modern currency). The American aviation production program was a dismal failure that inspired Congressional scrutiny in the immediate postwar period. The European emphasis on handicraft production facilitated rapid changes in design, which was crucial for survival in the air.

Foreign aircraft or not, the American aviators that made it to Europe made an important contribution to combat operations. Naval aviators in American designed and manufactured Curtiss flying boats searched for and fought German submarines off the coasts of England, France, Ireland, and Italy in the first operational missions of the war during the fall of 1917. Beginning in February 1918, marine bombing units began operations with RAF units in support of British and Belgian ground forces. The army air service began operations in France in June 1918 and its fighter, observation, bombardment, and balloon units quickly became involved in heavy fighting against the Germans. Captain Edward "Eddie" Rickenbacker (1890–1973), a famous prewar race-car driver who talked his way into flight training, became America's highest scoring ace with victories over twenty-one German aircraft and five observation balloons. He flew with the famous 94th Aero Squadron with its symbolic insignia of Uncle Sam's hat being thrown into the ring, or the fight, against Germany. The crews of the Liberty battle plane also fought heroically. Two air service lieutenants performed the first successful American military airlift operation when they dropped supplies to the famous "Lost Battalion" of the 77th Division as it fought to resist encirclement by the German army in October 1918, despite losing their lives in the process. Overall, the army's 1,500 aviators destroyed 781 German aircraft and 73 observation balloons at the cost of 583 casualties and 289 aircraft.

The last major Allied ground offensives of the war incorporated the major elements of airpower. American army Col. William "Billy" Mitchell (1879–1936) controlled a combined Allied air force of 1,500 aircraft during the reduction of the St. Mihiel salient, which had existed as a bulge in the Allied lines since the beginning of the war, in September 1918. He commanded the largest concentration of military aircraft up to that time and coordinated it with movements on the ground. Mitchell ordered 500 bombers and fighters to bomb and strafe German troops on the front line and to achieve air superiority. The remaining aircraft attacked troop assembly areas, supply routes, and communications networks right behind the lines to neutralize any German counterattacks. The Germans had already planned to withdraw from the area when Mitchell's aircraft attacked, but it was a valuable exercise in planning and doctrine. Later in September, Mitchell, promoted to brigadier general for his success, led a primarily American air force during the Meuse-Argonne offensive that continued until German capitulation in November 1918. As a result, Mitchell became America's most outstanding tactical air commander through his coordinated use of fighter, observation, and bombing aircraft.

THE GREAT WAR AND BEYOND

The airplane made dramatic strides during World War I in terms of both aeronautical technology and its application as a weapon of war. Before the war, many doubted the practicality of the airplane as a military weapon since it had been primarily a sensational form of entertainment. After the war, it was an accepted reality that had an unquestioned place in the world military establishment. Despite those gains, there was considerable skepticism over the specific roles airplanes would have in future wars. Douhet, Trenchard, and Mitchell would become the "prophets" of airpower, based on their experiences in the Great War, and their leadership and writings would influence and shape the use of the airplane in future wars. Airplanes were cheaper, easier to maintain, and evoked a sense of modern and futuristic technology that made them an attractive alternative to more-established technologies such as battleships. To them, airplanes made all other military technologies obsolete.

Despite the overall limited use of the airplane and its questionable influence on the outcome of the war, one thing was certain at the end of World War I: the airplane had reached an unprecedented position in twentieth-century popular culture. The romance, glamour, and action of aces locked in individual combat high above the trenches etched the airplane, and the pilot, into modern memory. Books from famous aces and aviation leaders such as Eddie Rickenbacker's *Fighting the Flying Circus* (1919), pulp novels and magazines, and Hollywood motion pictures such as *Wings* (1927), the first movie awarded an Oscar for best picture, Howard Hughes's blockbuster *Hell's Angels* (1930), and two versions of *The Dawn Patrol* (1930, 1938) reinforced that image. As late as the 1960s, many young boys and girls learned about the Great War through the exploits of one of their favorite comic-strip characters, Snoopy, from Charles M. Schulz's *Peanuts*. In *Snoopy and the Red Baron* (1966), the famous beagle, outfitted in his leather flying helmet, goggles, and scarf, was the great "World War I flying ace" that flew his Sopwith Camel, actually his doghouse, against the Red Baron in the skies over France.

3

The Aeronautical Revolution, 1918–1938

◆

On May 20–21, 1927, a small, silver airplane called the *Spirit of St. Louis*, with an unknown airmail pilot named Charles A. Lindbergh at the controls, flew 3,610 miles from Roosevelt Field, New York, across the Atlantic Ocean to Le Bourget Field outside of Paris, France, in 33½ hours. The Ryan-built airplane, named after Lindbergh's financial backers in St. Louis, combined some of the latest aeronautical developments—a radial air-cooled engine, an aluminum alloy propeller, a monoplane configuration, and long-range instrumentation—with some of those more established— fixed landing gear, a wood wing, and a tubular steel fuselage. Lindbergh's flight electrified the public with an intense fascination for aviation. The airplane and aviation in general was undergoing a significant transition at all levels in 1927. During the period 1918 to 1938, the unprecedented momentum created by parallel and intertwined advances in technology, governmental regulation, entrepreneurial growth, and cultural awareness culminated in an aeronautical revolution throughout Europe and North America.

DEFINING A ROLE FOR THE AIRPLANE, 1918–1925

The airplane emerged from World War I recognized widely for its military potential as a fighter, bomber, and observation platform. The war itself stimulated a chaotic period of limited growth that showed much promise to aviation enthusiasts. They believed airplanes would be valuable military and commercial technologies that would help usher in a new age characterized by life-improving technology. There was, however, still considerable doubt regarding the exact role of the airplane in the postwar world. The aeronautical community worked to define a role for the airplane and shaped aeronautical technology to meet that goal from 1918 to 1925.

In the early 1920s, a new generation of pilots, called barnstormers, trained during the war wandered the American countryside offering rides and performing aerial feats of skill and daring. The barnstormers bought surplus military aircraft left over from the Great War, primarily the Curtiss JN-4D Jenny trainer. The Jenny was a typical airplane of the time. It was easy to maintain and repair, but underpowered with its 90-horsepower Curtiss OX-5 V-8 engine. More important, these aircraft were sold at unbelievably low prices—sometimes as low as $50 each ($500 in modern currency)—on the civilian market. These aerial gypsies, as both individuals and performing together as flying circuses, gave the public its first exposure to the airplane and presented aviation as a grand spectacle with all of its glamour, excitement, and entertainment. In the process, the Jenny became one of the most famous airplanes in history. Barnstorming was, however, a blessing and a curse. It complemented the "ballyhoo" of the decade, but it made many people think that aviation would never amount to anything other than a form of entertainment.

The barnstormers brought the airplane to America's front door, but there was a significant undercurrent in aviation that was less well known. The national government became even more involved in the development of aviation. In essence, the U.S. government committed itself to making sure that the airplane would be a success and enthusiastically supported the new technology at technological, political, economic, and social levels. In support of that was a growing industry ready to use new technology to make aviation an important component of the national economy. The emergence of the modern airplane was a viable part of the American experience in the early 1930s.

While the military airplane rose to prominence during World War I, the idea of commercial aviation, or the aerial transportation of goods and services for profit, was still new. The Air Mail Service operated by the U.S.

The barnstormers flying aircraft, such as the Curtiss Jenny, performed many dramatic, and often deadly, stunts in front of excited crowds enthralled with the airplane. Historical Collection of Union Title Insurance and Trust Company, San Diego, California via National Air and Space Museum, Smithsonian Institution (SI A-337).

Post Office was an important precedent to the establishment of American commercial aviation during the period 1918–1925. The Post Office simply extended its scheduled land and sea operations into the air. The establishment of an all-weather, transcontinental, round-the-clock aerial mail delivery system from New York to San Francisco in 1921 pushed the boundaries of technology. The aircraft, heavily modified war-surplus DH-4 aircraft with strut-and-wire bracing, wood wings, tubular steel fuselages, and the liquid-cooled Liberty engine, flew in a variety of extreme climatic conditions and altitudes. More importantly, the organizational and technological infrastructure needed to keep the DH-4s in the air proved to be the key. The Air Mail Service began with a network of bonfires to mark the air routes, which quickly transformed into a sophisticated system of radio and light beacons. The fast delivery of mail aided American business through a route that took seventy-eight hours to travel. The National Aeronautics Association awarded the Air Mail Service the 1923 Collier Trophy, for the highest achievement in aviation for that year, for its consistent transcontinental and round-the-clock operation.

The Air Mail Service was very successful, but it did not support the manufacturers and air carriers within the American aviation industry. In both Europe and the United States, national governments recognized the need to support the young industry. In the United States, the year 1925 proved to be a watershed for commercial aviation. During the administration of President Calvin Coolidge, Congress turned over government-operated airmail routes to private carriers through postal contracts. The Air Mail Act of 1925 proved to be the beginning of the American commercial aviation industry. Rather than awarding contracts to the lowest bidder, the post office supported companies that exhibited a promise of long-term growth in both mail and passenger-carrying operations.

Commercial aviation was growing through the support of the post office, but the airplanes being used for the airmail were still very similar to those flown in World War I. The U.S. Army, Navy, and the National Advisory Committee for Aeronautics (NACA) as well as the manufacturers worked to change the shape, character, and use of the airplane. The American military attempted to gain a larger role in the immediate postwar period. Military leaders had to consider the role of the airplane as it pertained to traditional land and sea forces.

The reactions on the part of the army and the navy were similar and very different at the same time. The U.S. Army Air Service emerged from the war enthusiastic about the potential of airpower, the waging of war and the extension of a nation's political and military goals into the air. The average air service officer was typically young in age because it was a new branch of service within the army. The enthusiastic young aviators wanted a larger role for the airplane in the defense establishment. They emphasized that military aviation was the decisive force in warfare through the strategic bombing of enemy cities and industry as seen during World War I. As they pushed for their independence from the traditional land-based army, they set up a technological infrastructure that pushed aeronautical technology.

The air service conducted an aerial public relations campaign to prove the value of the military airplane and the need for an independent air force. Maj. Gen. Mason M. Patrick served as chief of army aviation from 1918 to 1927. Unlike his subordinates, he was an established career army officer with an engineering degree who learned to fly at the age of fifty-nine. Understanding the conservative nature of army bureaucracy, he quietly worked toward air force independence. In contrast to Patrick's tact and subtle advocacy, his subordinate, Brig. Gen. William "Billy" Mitchell, exhibited a flamboyant personality and exaggerated rhetoric that led to his 1925 court-martial for his outspoken views on what he believed to be a virtually nonexistent national aviation policy. Under their leadership, army aircraft sank and/or captured World War I German warships in 1920–1921 to prove that battleships were obsolete against airplanes, flew nonstop across the continental United States in 1923, flew around the world in 1924, and won the prestigious Schneider and Pulitzer racing trophies in 1925.

The army proved to be a major supporter of American aviation through the research and development activities conducted at its engineering facilities at McCook Field (1917–1927) and Wright Field (1927–1947) near Dayton, Ohio. Civilian and military engineers working for the army created new aeronautical innovations. For aircraft engines, superchargers and turbosuperchargers provided sea-level performance, where the air required for combustion is denser, at high altitudes, and high-octane fuels enabled more powerful engines with greater horsepower. The variable-pitch propeller allowed the angle, or pitch, at which each propeller blade rotated through the air to vary according to different flight conditions, which maximized the power of the aircraft engine much like an automobile transmission allowed a car to travel at different speeds. They also pioneered new engineering tools, including the army's first-ever wind tunnel, which was capable of speeds over 450 miles per hour. McCook Field was well known for its original aircraft designs, which were in competition with the struggling private manufacturing industry. When the time came to open Wright Field in 1927, the army utilized private contractors for new designs.

While the army aviators sought independence, naval aviators searched for ways to integrate aviation into the traditional seagoing navy. American naval aviation entered the immediate post–World War I period with 4,000 pilots and 30,000 enlisted personnel. In 1919, a large navy flying boat, the NC-4, became the first aircraft to cross the Atlantic Ocean by air. Naval aviators created a dedicated aviation service, called the Bureau of Aeronautics (BuAer) in 1921. The chief of the BuAer was Adm. William A. Moffett (1869–1933), who is considered the father of naval aviation. Moffett was an articulate and effective spokesman for naval aviation who worked to integrate the airplane into fleet operations and tactics. He emphasized a multirole mission for naval aviation. Building upon the already established roles of scouting and patrol with flying boats, Moffett worked to develop offensive uses such as dive-bombing and fighter pursuit.

The core of Moffett's work rested with the creation of a naval air force capable of sailing with the fleet anywhere in the world. The first American aircraft carrier, Langley, called the "covered wagon" for its resemblance to the prairie schooners of the nineteenth century, appeared in 1922 and served as the floating laboratory for naval flight operations and technology with its thirty-four aircraft. The knowledge gained from the Langley led to the creation of new specifications for aircraft designed specifically for naval operations. Designed by the Naval Aircraft Factory at the

Philadelphia Navy Yard, the TS-1 fighter was the first purpose-built American carrier airplane. For takeoffs from the short deck, the TS-1 relied upon a powerful 200-horsepower air-cooled Wright J-4 radial engine, which was light in weight and mechanically reliable. For landings, the rear fuselage of the TS-1 had a tail hook—found on all naval aircraft to this day—to engage arresting wires on the deck. Overall, these new aircraft had to be durable in a sea environment. The wood and fabric of the TS-1 deteriorated quickly at sea and the navy encouraged the use of metal as a structural material in both airframes and propellers in later aircraft.

The *Langley* experience also led to larger and faster aircraft carriers. The resultant *Lexington* and *Saratoga* appeared in 1927. These ships were capable of speeds up to thirty-three knots, which was enough for the carriers to produce their own wind for flight operations. During a simulated attack on the strategically important Panama Canal in 1929, carrier advocates effectively demonstrated the value of the aircraft carrier to naval operations. By 1941, the navy had eight operational aircraft carriers, 5,260 aircraft, 6,750 pilots, and 21,678 enlisted personnel.

With the American aviation industry in the doldrums in the early post–World War I period and the early 1920s, government support in the form of postal contracts, regulatory legislation, and commercial and military aircraft contracts fostered a growing manufacturing community found across the nation. On the East Coast, there was Curtiss on Long Island, Vought and Sikorsky in Connecticut, and Martin in Cleveland. On the West Coast, Boeing endured rainy Seattle while Douglas and Lockheed enjoyed an ideal climate for year-round flying weather in southern California. All of these companies bore the name of their founding members, who were aviation pioneers in their own right. The Wright Aeronautical Corporation produced Whirlwind radial engines in Paterson, New Jersey. Former Wright Aeronautical employees, under the leadership of Frederick B. Rentschler (1887–1956), formed Pratt & Whitney Aircraft and produced Wasp radial engines in Hartford, Connecticut. The nation's leading propeller manufacturers, Standard Steel Propeller and Hamilton Aero Manufacturing, were in Pittsburgh and Milwaukee, respectively.

As the American aviation industry grew, the NACA quietly worked behind the scenes to improve the performance of all aircraft in the area of aerodynamics and specifically in the use of streamline design. In the years immediately following World War I, it became increasingly clear that innovations in aerodynamics, primarily streamlining, would provide the efficiency needed for higher performance. The removal or redesign of drag-inducing structures such as landing gear, engines and radiators, struts and wires, open cockpits, and armament could greatly increase the performance of both civilian and military aircraft. In 1922, Louis-Charles Breguet (1880–1955), a pioneer aircraft manufacturer in France, issued a call for streamlined airplanes in his address, "Aerodynamic Efficiency and the Reduction of Aircraft Costs," presented before the Royal Aeronautical Society in London. To Breguet, improving an airplane's lift-to-drag ratio, what he called "fineness," through streamlining, was a technical challenge that would yield economic benefits by extending the overall operational range of an airplane. He suggested the use of new innovations such as retractable landing gear and went so far as to suggest an ideal fineness ratio, which would not be achieved until ten years later in the early 1930s.

CREATION OF THE MODERN AIRPLANE, 1926–1934

Airplanes and aviation in general underwent a significant transformation during the period 1926 to 1934. The widespread introduction of the revolutionary aeronautical innovations resulted in the emergence of the "modern" airplane, which possessed the major characteristics we associate with aircraft today. On a larger scale, modern aviation took shape at the political, economic, technological, and social levels.

While the Air Mail Act provided an opportunity for growth, commercial and private aviation required regulation to make it an overall safe activity. President Coolidge appointed Wall Street financier Dwight Morrow to chair a special board to investigate the current status of American aviation. The Morrow Board recommended the creation of a special aeronautics branch within the Commerce Department to regulate and foster civil aviation. Congress enacted the Air Commerce Act of 1926, which created the aeronautics branch of the Department of Commerce. The new organization had many important duties, including the regulation of pilots, aircraft, and operating procedures with safety as the primary issue, the maintenance and expansion of the lighted airways and radio navigation network, and the development of three transcontinental air routes to foster both passenger and airmail service. Finally, the aeronautics branch served as a booster for aviation by sponsoring competitions and programs that facilitated the expansion of aerial transportation.

The Air Commerce Act and Air Mail Act of the previous year stabilized the air routes and formed the nucleus of an American commercial aviation system with new airlines like American, Delta, and United. It became increasingly clear that government and industrial cooperation eliminated destructive competition and stabilized air transport. As the government took a larger role in supporting these new businesses, it also took a larger regulatory and promotional role.

The U.S. government provided sweeping change for military aviation as well. After the tumultuous court-martial and the public airing of grievances of Billy Mitchell, Congress created an autonomous air force within the army through the Air Corps Act of 1926. It was not full independence for American military aviation because the arrangement was similar to how the Marine Corps operated within the navy. With the organizational change came a five-year plan for expanding personnel and aircraft procurement, which allowed improved cooperation between the army and the industry.

While development of both commercial and military aviation at the government level continued in 1926, an obscure airmail pilot flying the St. Louis-to-Chicago route named Charles A. Lindbergh (1902–1974) began planning a solo transatlantic flight from New York to Paris. He wanted to win the much-publicized $25,000 Orteig Prize for the first nonstop flight from New York to Paris. He also saw it as an opportunity to showcase the capability of aeronautical technology by making a flight of more than 3,000 miles. Based in St. Louis, Lindbergh found support for his project from a group of entrepreneurs eager to support the enthusiastic young pilot. The end result of that partnership was an unprecedented cultural and financial response to the airplane.

Lindbergh possessed a visionary, but practical, understanding of 1920s aeronautical technology where he balanced the four systems of flight to create an overall successful long-distance airplane. His ideal aircraft was a single-seat monoplane powered by a single air-cooled Wright Whirlwind J-5C radial engine and a Standard Steel aluminum alloy propeller and capable of carrying more than 400 gallons of fuel. Unable to purchase that type of airplane, he contracted Ryan Airlines of San Diego, California, in February 1927 to build the design. The small company, known for its successful series of single-engine airmail aircraft with monoplane wings, employed thirty-five men and women in a former fish cannery on the San Diego waterfront. Lindbergh and Donald Hall, Ryan's chief engineer, focused on long-range and economical performance that maintained a balance between a myriad of design considerations, primarily long-range efficiency and safety.

Ryan completed the NYP (for *New York-to-Paris*) in April 1927. Lindbergh named it *Spirit of St. Louis* in honor of his benefactors back in Missouri. The new airplane featured the wood monoplane wing mounted on top of the fuselage, tubular steel fuselage, and fixed landing gear of other Ryan designs, but the NYP was much larger, with a wingspan of 46 feet, a length of 27 feet 7 inches, and an overall weight of 2,150 pounds. In addition to the Whirlwind engine and the Standard Steel propeller, other differences included the placement of the cockpit behind the gasoline and oil tanks for protection during a crash and the most advanced long-range instrumentation available, including an earth inductor compass. The design of the NYP was a stunning example of aeronautical technology in transition and was designed to meet one goal: flying the Atlantic nonstop.

Lindbergh was not the only individual trying to win the Orteig Prize. His transatlantic competitors, who included French World War I aces, American military and naval aviators, and notable civilian pilots, were all preparing for their own flights. Up to that point, none of them had been successful. Lindbergh left San Diego for New York by way of St. Louis on May 10, 1927, and set a new transcontinental record on the way. He was not able to leave for Paris until the morning of May 20, 1927, due to rain delays. Lindbergh carried with him four sandwiches, two canteens of water, and some prepackaged army rations to sustain him during the 3,610-mile flight that took more than thirty-three hours to complete.

The success of Lindbergh's flight contributed to the advancement of aviation in the United States and exhibited the growing superiority of American aviation technology. The previously unknown pilot was a true celebrity with the affectionate nicknames "Lucky Lindy" or the "Lone Eagle." Americans regarded him as a hero that represented all of the qualities that made the nation great: intellect, mechanical aptitude, determination, and strong will. Fascination with Lindbergh appeared at all levels

The Ryan NYP *Spirit of St. Louis* flown by Charles A. Lindbergh became both a technological and a cultural icon. Historical Collection of Union Title Insurance and Trust Company, San Diego, California via National Air and Space Museum, Smithsonian Institution (SI 79-14815).

of society and spurned new social activities such as the dance, the Lindy Hop. Lindbergh's retelling of the flight, *WE* (1927), which described his strong connection with the *Spirit*, was a national best seller and went through several editions. The response to Lindbergh represented trends that had been in motion even before he first contemplated his great achievement. There had been an overwhelming cultural response to the airplane well before 1927. The music, cinema, media, and print making up the popular culture of the time promised the arrival of a new age brought by the progress of technology. To believe in the airplane and what it could do for humankind meant to be "air-minded" in spirit.

With a nation already enthralled with the airplane, the excitement created by Lindbergh's flight contributed to greater confidence in aviation on Wall Street, the financial center of the United States. Quickly, consolidation of the pioneer aviation companies through merger and acquisition resulted in the creation of large holding companies that represented a "full-service" approach to manufacturing, selling, and transportation within the entire aviation industry. Two major corporate entities emerged. The first, and by far the largest, was the United Aircraft and Transport Corporation, which produced Boeing, Vought, and Sikorsky airplanes, Pratt & Whitney engines, and Hamilton Standard propellers (after the merger of those two companies). United's competitor, Curtiss-Wright, represented two of the oldest names in American aviation.

Less than three months after Lindbergh's flight, on July 4, a revolutionary new type of airplane from Lockheed, called the Vega, took to the air. It resembled Lindbergh's *Spirit of St. Louis* with its monoplane wing mounted atop the fuselage, radial engine, and fixed landing gear, but the two aircraft differed greatly from that point. The Vega featured a circular monocoque plywood stressed-skin fuselage (like the Deperdussin racer before the Great War) and a cantilever wing. The absence of external struts and wires reflected the intention of its designer, John K. "Jack" Northrop (1895–1981), for the aircraft to be as streamline, or what he called "clean," as possible. Northrop was a self-taught, intuitive engineer capable of successfully transferring the three-dimensional ideas for aircraft in his mind into reality. His elegantly simple method of construction was both strong and aerodynamically efficient. Lockheed workers formed the plywood halves of the fuselage with two large concrete molds and glued them together to create a smooth and hollow structure. All of those innovations enabled the six-passenger Vega to achieve a maximum speed of 190 miles per hour.

The Vega was a dramatic departure in aeronautical engineering and the pioneering pilots of the day found it to be the perfect airplane for the great flights of the late 1920s and early 1930s that crossed oceans, continents, and technical boundaries. Amelia Earhart (1897–1937), known as Lady Lindbergh, achieved many firsts flying Vegas. She was the first woman to fly solo across the United States (May 1932), across the Atlantic (August 1932), and across the Pacific from Hawaii to California (January 1935). Building upon that success, Earhart endeavored to be the first person to fly around the world, but disappeared with her navigator, Fred Noonan, over the Pacific in July 1937.

Another notable Vega pilot was Wiley Post (1898–1935), a veteran of the Oklahoma oil fields who lost an eye in an accident and flew the white and blue *Winnie Mae*. After breaking a series of transcontinental records, he flew around the world with navigator Harold Gatty in 1931 in eight days. Two years later, Post flew solo around the world in only seven and one half days during the summer of 1933. He went on to pioneer high-altitude flight in the world's first pressure suit, which enabled him to comfortably fly at heights up to 50,000 feet. In the process, Post discovered the jet stream, the fast, westerly winds, which enabled him to exceed the regular speed of the Vega by more than 100 miles per hour.

Post's *Winnie Mae* was an improved model of the first Vega. In terms of streamlining, Northrop's design had two problem areas: the landing gear and the radial engine. With no area to retract the landing gear, Lockheed covered the wheels in teardrop-shaped streamline fairings, called "spats" after the shoe coverings men wore with formal dress, to reduce drag. Later Northrop designs, such as the Gamma fast airmail plane, would feature "pants," which covered the entire landing gear.

The Vega benefited greatly from the inclusion of a 450-horsepower Pratt & Whitney Wasp air-cooled radial engine, which was both powerful and light. Unfortunately, the protruding cylinders of the engine created a lot of drag for an otherwise streamline airplane. As the Vega and similar advanced radial engine-powered aircraft were taking to the air, pioneering work in aerodynamics conducted by the NACA resulted in a new circular radial engine cowling that reduced drag and improved cooling at the same time. At the NACA's Langley Memorial Aeronautical Laboratory near Hampton, Virginia, engineer Fred E. Weick (1899–1993) and his colleagues addressed the fundamental problem of incorporating a radial engine into aircraft design in a wind tunnel with a 20-foot opening that allowed them to test full-size aircraft structures. Their solution, called the NACA cowling, arrived at the right moment to increase the performance of new aircraft. The famous pilot, Frank Hawks (1897–1938), flew the Texaco Lockheed Air Express, with a NACA cowling installed, from Los Angeles to New York nonstop in a record time of 18 hours, 13 minutes in February 1929. Tests of a Curtiss AT-5A Hawk fighter with a NACA cowling increased its top speed by 16 percent. The contribution was so great to overall aircraft design that the NACA won its first Collier Trophy in 1929 for its innovative work.

The success of Jack Northrop's intuitive design and the NACA's systematic research were indications that the airplane was changing. One of the key ways to improve aircraft performance was aerodynamic drag reduction through streamlining. Great Britain's leading aerodynamicist, Dr. B. Melville Jones (1887–1975) of Cambridge University, presented a paper entitled "The Streamline Airplane" to the Royal Aeronautical Society in 1929. Jones outlined how streamlining reduced drag to generate higher cruising speeds and lower fuel consumption that resulted in increased range and payload. Noting the inefficiency of previous aircraft designs, the professor cited the specific decrease in form drag was a major obstacle to increased aerodynamic efficiency. The aeronautical community enthusiastically received Jones's landmark speech. It was the first major acknowledgment of the benefits of streamlining.

The first generation of university-trained aeronautical engineers began to enter industry, the government, and academia. Through the philanthropic Daniel Guggenheim Fund for the Promotion of Aeronautics, new aeronautical engineering schools, complete with wind tunnels, appeared at the California Institute of Technology, Georgia Institute of Technology, Massachusetts Institute of Technology, University of Michigan, New York University, Stanford University, and University of Washington. The creation of these dedicated academic programs ensured that aeronautics would be an institutionalized profession.

These intertwined trends came together in the early 1930s to create the modern airplane. New types of aircraft started incorporating streamline design, radial engines with NACA cowlings, variable-pitch propellers, retractable landing gear, and other new innovations, but they were used in a piecemeal fashion. Competition between two airlines resulted in the creation of one aircraft that would have all of these major advances. Jack Frye (1904–1959), vice president of Transcontinental and Western Air (TWA), needed a new airliner. The public considered TWA's old Fokker trimotor airliners to be dangerous because one of the aircraft crashed with the famous Notre Dame football coach, Knute Rockne, aboard in 1931. Investigators ruled that structural failure of the wooden wing was the cause. Frye knew that Boeing was designing a new all-metal, twin-engine, monoplane airliner that would give him the performance he wanted. Boeing and TWA's archrival, United Airlines, were part of the United Aircraft and Transport Corporation. None of the new Boeing airliners, called the 247, would be made available to TWA until long after United Airlines had all they wanted.

The exclusion of TWA from purchasing the 247 was a fortunate occurrence in the long run. Frye defiantly distributed a letter and one-page
specifications sheet for a new trimotor airliner design to the leading
American aircraft manufacturers—Curtiss-Wright, Ford, Martin, Consolidated, and Douglas—in August 1932. Douglas won the contract with an
all-metal, twin-engine, monoplane design capable of carrying twelve (and
later fourteen) passengers. The small Santa Monica, California, company
was well known for building the first airplanes to circumnavigate the earth
in 1924, the army's Douglas World Cruisers. The company's namesake,
Donald W. Douglas (1892–1981), was one of the first university-trained
aeronautical engineers in the United States. The engineers who would lead
the development program of what would be the first commercial product
for the company, the DC, or "Douglas Commercial," was Arthur E.
Raymond (1899–1999) and James H. "Dutch" Kindelberger (1895–1962).

The prototype Douglas airliner, called the DC-1, flew in July 1933. In
every sense of the word, it was a streamline airplane due to the extensive
amount of wind tunnel testing at Guggenheim Aeronautical Laboratory at
the California Institute of Technology (GALCIT) used in its design. The
production aircraft, the DC-2, quickly followed with its first flight in May
1934. The new airplane would have all of the major innovations of
the Aeronautical Revolution: an advanced NACA-designed airfoil, a high-
aspect ratio wing, radial engines with NACA cowlings, variable-pitch
propellers, retractable landing gear, and effective, but remarkably simple,
high-lift split flaps. The cantilever wing was a multicellular design from
Jack Northrop that could withstand a small steamroller going over it.
Overall, the DC-2 carried fourteen passengers while cruising at 212 miles
per hour.

In October 1934, Douglas DC-2 and Boeing Model 247 transports
finished second and third, respectively, behind a special-purpose British de
Havilland DH.88 Comet long-distance racer in the 11,300-mile England-
to-Australia MacRobertson Trophy Race. The *New York Times* hailed the
performance of the two revolutionary aircraft as a victory for American
aeronautical technology and its apparent superiority over European technologies. The much larger American transports, especially Royal Dutch
KLM's DC-2 *Uiver*, carried passengers and mail much like they would have
in regular commercial operations while the British entry carried only its
pilot and copilot. The modern airplane had arrived and airlines around the
world stood in line to get the new DC-2.

Airline competition once again stimulated the continued development
of the Douglas airliners. Cyrus R. Smith (1899–1990), president of
American Airlines, needed a new fleet of aircraft that would provide

The DC series of airliners, such as this American Airlines DST, embodied all the innovations that resulted in the "modern" airplane. National Air and Space Museum, Smithsonian Institution (SI 91-7082).

nighttime transcontinental service. His engineers told him that a DC-2 with the fuselage widened by 26 inches and a wingspan increased by 10 feet would allow fourteen sleeping berths. Smith placed a long-distance telephone call, unheard of at the time, to Douglas asking for a new airplane. The result, called the Douglas Sleeper Transport (DST), first flew on December 17, 1935, thirty-two years after the Wrights' flight at Kitty Hawk. The daytime passenger version, the DC-3, carried twenty-one people and became the most popular and reliable propeller-driven airliner in aviation history. The appearance of the DC-3, considered to be the first airplane to make money by carrying passengers and not mail, marked the culmination of the design revolution in American aircraft.

REVOLUTION AND REFINEMENT, 1935–1938

The Aeronautical Revolution witnessed the transformation of the slow, fabric-covered, strut-and-wire-braced biplane of 1918 into the modern high-speed, streamline, cantilever-wing monoplane of 1938. The aeronautical community endeavored to refine those advances in technology and infrastructure during the period 1935 to 1938 to increase further the performance of the airplane and its role in modern life. In many ways, what we consider to be "modern" aviation in terms of technology, economics, politics, and war began during the second half of the 1930s.

With the basic formula for the modern airplane in place, the aeronautical community began to push the limits of conventional aircraft design. The NACA, building upon its success with the cowling research, concentrated on the aerodynamic testing of full-scale aircraft in wind tunnels. The Full-Scale Tunnel (FST), with a 30-by-60-foot test section, opened at Langley in 1931. The building was a massive structure 434 feet long, more than 200 feet wide, and nine stories high. The first aircraft to be tested in the FST was a navy Vought O3U-1 Corsair observation airplane. Quickly, the testing began to focus on removing as much drag from an airplane in flight as possible. NACA engineers, through an extensive program involving the navy's first monoplane fighter, Brewster XF2A Buffalo, showed that attention to details such as air intakes, exhaust pipes, and gun ports effectively reduced drag.

American military aviation made strides toward a strategic air force. The Air Corps created the General Headquarters (GHQ) Air Force in 1935 to concentrate solely on developing the tactics and technology required to wage strategic warfare. Those two areas came together in the concept of daylight, high-altitude, precision bombing. By attacking an enemy's ability to wage war through the destruction of its industrial production and infrastructure, the GHQ Air Force aimed to be a deterrent to large-scale conflict in the world. The delivery system was fast, high flying, multi-engine aircraft such as the Boeing B-17 Flying Fortress, which embodied the latest innovations of the Aeronautical Revolution and a new military weapon, the Norden bombsight, allowing them to destroy specific military/strategic targets while at the same time easily out-flying any enemy aircraft attempting to intercept them. The Air Corps believed strongly in the potential of the airplane in winning future wars. By the end of the decade, the leading nations of Asia, Europe, and North America would find themselves fighting each other on a world scale with airplanes.

4

World War II in the Air, 1939–1945

In August 1945, World War II, the bloodiest conflict in human history, ended in the skies over Japan. Over the course of two separate days, the crews of two Boeing B-29 Superfortress bombers, named the *Enola Gay* and *Bock's Car*, dropped atomic bombs on the cities of Hiroshima and Nagasaki. These new weapons obliterated almost 100,000 people in an instant. The delivery system, the B-29, was the most advanced propeller-driven airplane in the world with aerodynamic streamline design, all-metal monoplane construction, four high-horsepower engines, retractable landing gear, a pressurized cabin, and an advanced defensive weapon system. From 1939 to 1945, World War II raged across the major continents and oceans. Airpower became a major component of the military establishment during World War II, when the airplane became a decisive weapon in modern warfare. The rise of the military airplane came with a large cost. The economic, financial, and technical resources of entire nations were needed to produce new bombers and fighters. With the increased use of the airplane came the realization that the specific targeting and bombing of civilian populations had become an accepted fact.

The crew of the Boeing B-29 Superfortress *Enola Gay* returns after dropping the first atomic bomb in history on Hiroshima, Japan. U.S. Air Force via National Air and Space Museum, Smithsonian Institution (SI 2000-4554).

THE RISE OF AIRPOWER

After the bloodbath of World War I, the world's leading nations aimed to avoid future wars through diplomatic, political, and even technological solutions. The rise of authoritarian governments in Europe and Asia in the 1920s and 1930s, however, made war almost inevitable. Nazi Germany, under the leadership of Adolf Hitler, secretly rebuilt its military to include a world-class air force in direct violation of the Versailles Treaty. Benito Mussolini's fascist Italy joined Germany with the intention of dominating Europe and Africa. Great Britain and France, in alliance with smaller nations in Europe, prepared to meet the threat. On the other side of the world, Imperial Japan began to spread its influence into Manchuria and China with the potential of further expansion into the rest of Asia and the Pacific in direct conflict with American and European interests. As a result of these tensions, the major nations of Europe, North America, and Asia mobilized their war industries in the late 1930s, which included the production of new aerial weapons of war as a primary focus.

There was an important prelude to World War II in the air during the Spanish civil war (1936–1939). The bitter campaigns fought between the Nationalist forces, supported by Germany and Italy, and the Republicans, receiving aid from the communist Soviet Union, for control of Spain included a significant use of airpower. The newly revived German air force, the Luftwaffe, used the conflict to develop the tactics and technology it would use in future campaigns. German Junkers Ju-52 transports carried

National troops from Spanish Morocco to Spain itself to start the war. Luftwaffe pilots in German-built fighters fought for air superiority against Republican pilots in Soviet-built fighters. The most infamous aerial episode of the war was the destruction of the Basque town of Guernica by German bombers in April 1937. For two hours, waves of bombers and fighters bombed and strafed the city, effectively destroying the city and killing hundreds of civilians. The modern artist Pablo Picasso (1881–1973), believed to be one of the best artists of the twentieth century, memorialized the destruction in his mural *Guernica* (1937), which proved to be a poignant symbol of the horror of war throughout the course of the twentieth century.

After expanding as much as it could without going to war, Germany started World War II with the invasion of Poland on September 1, 1939, in a highly coordinated tactical air-land offensive called blitzkrieg, or "lightning war." German infantry and tank units moved fast across a battlefield. If they encountered a pocket of resistance, they would encircle it and keep moving forward. Overhead, Luftwaffe bomber and fighter aircraft protected the advance, attacked obstacles such as tanks, fortifications, and encircled enemy units, and destroyed any opposing air forces on the ground and in the air. The Luftwaffe's aircraft were especially suited for blitzkrieg. The Junkers Ju-87 Stuka dive-bomber was capable of hitting specific targets accurately and had a wind-driven siren mounted on its landing gear that increased the psychological impact of an attack, especially on fleeing civilians. The Heinkel He 111 twin-engine bomber carried enough bombs to knock out troop concentrations and supply lines. The fast and nimble Messerschmitt Bf109 fighter could out-fly and outshoot anything in the air. By the spring of 1940, the Luftwaffe had destroyed the Polish Lotnictwo Wojskowe, the French l'Armée de l'Air, and the British Royal Air Force units sent to fight on the continent.

The fall of France left Great Britain alone to face the expansion of Nazi Germany. The German military formulated a plan for an invasion called Operation Sea Lion to capture the British Isles. Before the invasion fleet could set sail across the English Channel, the Luftwaffe had to achieve air superiority by destroying the Royal Air Force before the blitzkrieg could begin. What resulted was one of the most dramatic campaigns of the war, the Battle of Britain. The standoff between the Royal Air Force and the Luftwaffe during the summer of 1940 highlighted the importance of the fighter airplane to overall war strategy and the differences between the two countries on their use and employment of airpower technology.

The Luftwaffe had more than 2,000 frontline fighters and bombers for the campaign, but the aircraft and tactics that made blitzkrieg possible did not ensure that Germany would easily defeat England. The small, single-seat, low-wing Messerschmitt Bf109 monoplane featured a heavy cannon and machine gun armament, was highly maneuverable, and was the preferred mount of most German aces. The Bf109 was the product of one of the most famous and talented German aircraft designers, Professor Wilhelm "Willy" Messerschmitt (1898–1978). The Bf109 had one of the most advanced aircraft engines in the world, the 1,400-horsepower Daimler-Benz DB 600 inverted V-12, liquid-cooled engine. Rather than having the cylinders sit atop the crankcase, the DB series had them at the bottom, which provided better visibility for the pilot, enabled better access for maintenance, and lowered the aircraft's center of gravity for better performance. The DB 600 fuel injection allowed pilots to put the Bf109 in extreme negative g-force maneuvers that forced engines equipped with gravity-fed carburetors to cut out during aerial combat, an occurrence no fighter pilot wanted to experience. Despite its popularity and potency as a deadly fighter, the Bf109 had a severely limited operational range. It did not have the fuel capacity to enable it to spend much time over England due to the distance of its airfields located in France.

RAF Fighter Command, under the leadership of Air Marshal Sir Hugh Dowding (1882–1970), had a sophisticated chain of radar stations and ground-based observers, a radio-based network of aircraft direction and control, and approximately 800 fighters to oppose the Luftwaffe. Radar, short for *radio detection and ranging*, was a new technology that used radio waves to detect flying aircraft that gave the British advance notice and the location of German bombing raids. A veritable army of ground-based observers kept track of the raids once they reached England. Controllers, housed deep within bombproof bunkers, took that information to communicate with and direct individual RAF squadrons to their targets by radio. As a result, a RAF fighter squadron could be in the air ready to fight the Luftwaffe within five minutes. This efficient and integrated air defense system allowed the smaller number of RAF fighters to attack the numerically superior German bombers and fighters very effectively.

The RAF's primary fighters were the Hawker Hurricane and Supermarine Spitfire. Both evolved from a 1934 Air Ministry specification for a defensive fighter monoplane with eight machine guns, an enclosed cockpit, retractable landing gear, and a Rolls-Royce liquid-cooled twelve-cylinder vee engine. The Hurricane was the RAF's first monoplane fighter to exceed 300 miles per hour in level flight when RAF squadrons began the transition from biplane fighters in December 1937. Its designer, Sydney Camm (1893–1966), used a conservative engineering approach. The metal and wood fuselage and fabric covering facilitated production of the aircraft in large numbers. By 1939, the Hawker fighter was slower than and not as maneuverable as the Bf109. As a result, Fighter Command directed Hurricane squadrons to destroy Luftwaffe bomber aircraft. Many of the RAF units flying Hurricanes were composed of pilots from the nations that fell to the blitzkrieg, primarily Czechoslovakia and Poland.

The RAF's use of the Supermarine Spitfire during the Battle of Britain has become legendary. The Spitfire was the state of the art in fighter design when it entered operational service in 1938. Supermarine's chief engineer and designer, the self-taught Reginald J. Mitchell (1895–1937), and his colleagues used the experience of designing aircraft for the international Schneider Trophy racing competition during the 1920s to design the new fighter. French industrialist and early aviator Jacques P. Schneider (1879– 1928) created the Coupe d'Aviation Maritime Jacques Schneider in 1913 to encourage the development of commercial seaplanes. Teams from Italy, France, and the United States competed for the prestige of winning the trophy in what were then the world's fastest airplanes. RAF Flt. Lt. John N. Boothman (1901–1957) flew a Supermarine S.6B seaplane racer at almost 5½ miles per minute (340 miles per hour) to win the final Schneider Trophy competition in front of thousands of excited spectators at Calshot, England, in September 1931. When Great Britain needed a modern fighter in the mid-1930s, Mitchell had the formula for a high-performance airplane with all-metal monocoque construction, esthetically pleasing elliptical wings that reduced drag, and speed capable of approaching 400 miles per hour.

At the outbreak of the war, the Spitfire and the Bf 109 were evenly matched. The British fighter was faster and more maneuverable, but the German airplane could climb and dive better than its British adversary. The major disadvantage for the RAF, however, was that it did not have enough Spitfires.

Both the Spitfire and the Hurricane used the same engine, the Rolls-Royce Merlin. The Merlin was a crowning achievement in a highly successful line of aircraft engines produced by Rolls-Royce. In 1904, Charles Stuart Rolls (1877–1910), a prominent automobile and aviation enthusiast, and Frederick Henry Royce (1863–1933), a highly creative engineer, formed a partnership to manufacture world-class luxury cars. After Rolls's death in a flying accident, Royce continued with the company and began designing and building aircraft engines during World War I for the British government. After achieving great success with the R series for the Schneider Trophy racers, Royce began development of a new high-performance engine in 1933. The Merlin was a dramatic example of progress in aircraft engine technology during the years between the two

Legends with elliptical wings: three RAF Supermarine Spitfires during the Battle of Britain. National Air and Space Museum, Smithsonian Institution (SI 93-5362).

world wars. When equipped with a supercharger, the Merlin was capable of generating 1,000 to 2,000 horsepower. It generated approximately six times more horsepower than the Liberty V–12 engine of World War I and became just as famous as the Spitfire and the other aircraft it powered.

The Battle of Britain raged from July to October 1940. The Luftwaffe designated the RAF as its primary strategic target in preparation for the invasion. The German bombers brought the RAF close to an operational collapse. Hitler and his Luftwaffe chief, Hermann Göring (1893–1946), expanded the available targets to focus on British cities and civilians. That decision made the raids, called the Blitz by the British, the first major campaign of the war to target civilians and gave the RAF a much-needed break from direct Luftwaffe attacks. The quick blurring between what constituted a military and civilian target also highlighted the fact that the Luftwaffe's Heinkel He 111 and Junkers Stuka bombers did not possess the speed, range, and payload that would make them adequate strategic bombers. Luckily for the British, the RAF air defense system allowed the numerically inferior fighter squadrons to fight an effective defensive campaign against the Luftwaffe. In the end, the German military just gave up its plans to invade Great Britain because they could not control the skies overhead. German bombers would continue to attack British cities for the remainder of the war, but it would not be in preparation for an invasion.

The RAF victory in the Battle of Britain became a legendary tale of a "valiant few" that saved the island nation. British cinema celebrated the triumph as early as 1942 in the film *Spitfire* where Mitchell and his fighter had equal parts. Almost thirty years later, the 1969 epic *Battle of Britain* told the story from the view of both the British and the Germans.

AMERICA ENTERS THE WAR

As the British fought back the Nazi blitzkrieg, the Imperial Japanese Empire, undeterred by American political and economic resistance to its successful military expansion into Asia, decided to expand its sphere of influence across the Pacific Ocean. The officer in charge of the new war plans, Japanese Navy Admiral Isoruku Yamamoto (1884–1943), reasoned that a strategic strike against the U.S. fleet at anchorage at Pearl Harbor in the Hawaiian Islands with carrier aircraft coupled with the capture of American, British, and Dutch possessions in the Pacific would consolidate Japan's military and diplomatic position in the world community. The admiral believed that such an attack—based upon the successful British Royal Navy surprise attack on the Italian fleet at Taranto in 1940—would give Japan at least six months to prepare for a revitalized United States to go on the offensive.

The Imperial Japanese Navy possessed one of the world's best carrier forces in 1941. The Mitsubishi A6M fighter, better known as the Zero, was a world-class design that was fast and highly maneuverable, and could fly distances up to 2,000 miles. To achieve such stunning performance, engineer Jiro Horikoshi (1904–1982) had to forgo adequate defensive armor, incorporate an overly lightweight fuselage and wing construction, and use a low-horsepower radial engine. The Aichi D3A Val dive-bomber looked antiquated with its fixed landing gear, but it was deadly accurate against ships at sea. The Nakajima B5N Kate could deliver a bomb from a high altitude or drop a torpedo at sea level to destroy its target.

The attack on Pearl Harbor on December 7, 1941, was a masterful combination of planning, strategy, and luck. Six Japanese carriers under the direct command of Admiral Chuichi Nagumo (1887–1944) launched more than 300 Kate, Val, and Zero aircraft in two separate waves. That caught the Americans completely by surprise. The attackers achieved air superiority by neutralizing the army, navy, and marine airfields and then disabled or sank eight battleships, including the fleet flagship *Arizona*, and ten other vessels, and inflicted more than 2,000 casualties. The Japanese raiders left the strategically important ship repair facilities and fuel tanks unharmed and missed the aircraft carriers, which were out to sea.

Almost immediately, the Japanese attack on Pearl Harbor proved the importance of carrier aviation tactics in overall war strategy. It also galvanized American public opinion in favor of entering World War II on the Allied side against both Japan and Germany, after the latter declared war soon after the attack. Soon the phrase "Remember Pearl Harbor" echoed throughout all levels of society as America prepared for a global war in the air. Despite that enthusiasm, America had to fight a defensive war until it could fully mobilize its military war machine.

The first Americans to fight the Japanese were a small group of pilots fighting for Nationalist China called the American Volunteer Group (AVG) and better known as the Flying Tigers. Led by a retired army air corps specialist in fighter tactics, Claire Chennault (1893–1958), the 300 former army, navy, marine, and industry pilots and ground personnel defended China's last main supply route to the outside world, the Burma Road, in the face of overwhelming Japanese air and ground forces. The AVG operated three squadrons—the Hell's Angels, Panda Bears, and the Adam and Eves—against the Japanese beginning in December 1941. Over the course of seven months, the independent group of volunteers destroyed approximately 300 Japanese aircraft. When the Burma Road fell to the Japanese in May 1942, the group disbanded the following July and became the U.S. Army's 23rd Fighter Group. The AVG became world famous during the dark early days of American entry into World War II through constant exposure in the media and through the widely popular John Wayne film *Flying Tigers* (1942).

Like the Spitfire during the Battle of Britain, the AVG's Curtiss P-40 Warhawks, with their distinctive shark's teeth insignia, became legendary. Unlike the Spitfire, however, the P-40 was an outdated design in 1941 even though it was the American Army's frontline fighter at the outbreak of the war. Curtiss designer Donovan R. Berlin designed an all-metal, monoplane fighter powered by a radial engine called the P-36 Hawk. The addition of a 1,150-horsepower liquid-cooled Allison in-line V-12 engine and more machine guns resulted in the P-40, which first flew in October 1938. The Warhawk was unable to outmaneuver the more advanced German and Japanese fighters in combat. Nevertheless, it was the only fighter available in large numbers for the United States and its allies, including France, Great Britain, the Soviet Union, and Nationalist China, until higher performance aircraft became available.

The first official American military action taken against the Japanese occurred in April 1942. The United States intended to attack the primary Japanese ports, Tokyo, the capital, and Yokohama. Well-known aviation hero Col. James "Jimmy" Doolittle (1896–1993) was to lead a formation of specially trained crews in sixteen North American B-25 Mitchell medium bombers (named after Air Force pioneer Billy Mitchell) from the aircraft carrier *Hornet*. After the raid, the bombers were to land at friendly airfields on the Chinese mainland. Fearing early detection by the Japanese, the B-25 pilots launched early and attacked their targets, but all were forced to crash-land due to lack of fuel. Most of the crews escaped to return to the United States, but a few were captured, tried, and executed by the Japanese for their participation in the raid.

Despite the fact that it produced little actual damage to the Japanese war effort, the Doolittle raid was a much-needed morale boost for the United States in the dark early days of World War II. It elevated Doolittle even higher as an American hero when he received the Congressional Medal of Honor for his actions. One of the best books (1943) and subsequent films (1944) from the war era, *Thirty Seconds Over Tokyo*, brought to the American public the story of the raid through the experience of the author and raider, Ted Lawson. The raid also influenced the Japanese to alter its strategy in the Pacific by identifying the small American outpost on Midway Island, located approximately in the center of the Pacific, as a major objective.

Attempting to complete their timetable of conquest, the Japanese planned to capture Port Moresby in New Guinea to achieve a strategic location for further expansion into the southwest Pacific. American resistance to the move, informed by intelligence, resulted in the Battle of Coral Sea in May 1942, the first naval engagement in history fought entirely by carrier-borne aircraft. The battle was a stalemate that was technically in favor of the Japanese, who lost many experienced pilots and aircrew and one light carrier in exchange for the American loss of the carrier *Lexington* and the severe damaging of the *Yorktown*. For the Americans, however, it was a strategic victory because it stopped Japanese expansion into Australia.

Undeterred by the stalemate at Coral Sea, the Japanese fleet concentrated on the capture of Midway Island. Admiral Yamamoto intended to capture Midway and, in the process, draw out the remaining American carrier forces and destroy them once and for all with six carriers and more than 400 aircraft. Once again, American intelligence deciphered the Japanese code and revealed Yamamoto's plans in enough time for Adm. Chester W. Nimitz (1885–1966) and the U.S. Pacific Fleet to stage an ambush with three carriers and little over 200 aircraft in June 1942.

Through a combination of luck, skill, and tactics based on the capabilities of the Douglas SBD Dauntless dive-bomber, the Americans destroyed four Japanese carriers, more than 200 aircraft, and most of the Imperial Navy's highly skilled pilots in return for the loss of the *Yorktown*. The battle of Midway became the turning point of the war in the Pacific.

ALLIED AIRPOWER ON THE OFFENSIVE

With the European continent occupied by the Nazis and the fate of North Africa and the Soviet Union in doubt, American President Franklin D. Roosevelt and British Prime Minister Winston Churchill agreed at the Casablanca Conference in January 1943 to adopt a strategy of "bombing around-the-clock." With German industrial, communications, transportation, and power networks as the target, the Royal Air Force would bomb by night and the American Army Air Forces would bomb by day in a combined offensive. In June 1941, the army's air corps became the army air forces, an autonomous service within the larger army with Gen. Henry H. "Hap" Arnold (1886–1950) as its leader. Under the new organization, a specific combat air force, consisting of both fighter and bomber units and designated a number for identification, would operate in a theater of operations. The Eighth Air Force based in England and the Fifteenth Air Force in Italy would bring the American doctrine of daylight, high-altitude, and precision bombing to the skies over Nazi Germany.

The technological foundation for American strategic bombing doctrine rested on two weapons: a four-engine airplane able to carry 2,000 pounds of bombs and a bombsight to make sure they hit the target. The army air forces used two long-range, four-engine bombers, the Boeing B-17 Flying Fortress and the Consolidated B-24 Liberator, in Europe. Each aircraft, powered by high-horsepower radial engines equipped with supercharging, carried roughly a ten-man crew. The pilot, copilot, and navigator guided the aircraft to and from the target. The radio operator maintained communication with the outside world. The aircraft crew chief, responsible for the mechanical operation of the bomber, operated a gun turret behind the cockpit. Dedicated defensive gunners underneath, on each side, and at the tail of the aircraft protected the aircraft from German interceptors.

The goal of those ten crew members was to get the bomber to the target. Once there, the bombardier had the job of directing the load of bombs to the target. The bombardier's tool was the Norden bombsight, an elegant analog computer that calculated bomb trajectory. Army theorists believed that Norden-equipped bombers could deliver destruction accurately to an enemy's ability to wage war without unnecessary loss of civilian lives. The American government considered the mechanical workings held within its black metal casing to be an official state secret and guarded it cautiously throughout the war. A real Norden bombsight could not even appear in the Hollywood film *Bombardier* (1943), which celebrated it and the young men who trained to use it.

The Eighth conducted its first bombing raid on targets in German-occupied France beginning with Rouen in August 1942. The following October, more than 100 B-17s and B-24s bombed the town of Lille. These raids allowed the Eighth's bomber units to gain operational experience under the protection of Republic P-47 Thunderbolt and Lockheed P-38 Lightning escort fighters. The limited range of these two fighters, both of which became famous over the course of the war, prevented them from venturing far into German territory.

When the Eighth began to penetrate further into Germany and beyond the range of escorts, the price of daylight operations became horribly apparent. American air leaders believed that heavily armed B-17s and B-24s flying at high altitude in defensive "combat box" formations could protect themselves from German fighters. Raids on important German targets proved otherwise. During one week in October 1943, the Eighth lost 150 bombers to German fighters and antiaircraft artillery, called "flak." When looking at that statistic, it must be remembered that one bomber carried a ten-person crew, which meant the Eighth lost 1,500 men. Not able to withstand such losses, Eighth Air Force leader, Gen. Ira C. Eaker (1896–1987), ordered a halt to all attacks beyond the range of escort fighters. The Luftwaffe had retained air superiority over Germany.

What the army air forces needed was an escort fighter capable of going deep into German territory with the bombers. The introduction of the North American P-51 Mustang in February 1944 renewed the daylight bombing offensive. The Mustang's powerful 1,650-horsepower Merlin engine enabled it to outperform any German fighter and destroy them with its six wing-mounted machine guns. Equipped with external fuel tanks that could be dropped when needed, the P-51 had an extended range of 850 miles, enough to fly to Berlin and back.

With the new airplane came new tactics. While American bombers attacked German targets, the P-51 units would seek out and destroy the Luftwaffe's fighter force. The African American pilots of the 332nd Fighter Group, called the Red Tails due to the distinctive markings on their Mustangs, operated with the Fifteenth Air Force in Italy. Despite cultural and institutional resistance to their participation in the war, these highly

P-51 Mustang fighters, equipped with long-range drop tanks under each wing, were key to the success of the strategic air war over Europe. U.S. Air Force via National Air and Space Museum, Smithsonian Institution (USAF-52830AC).

proficient pilots never lost a bomber under their protection. The Red Tails were part of a larger group of African American pilots and ground personnel called the Tuskegee Airmen after their training facility in Alabama.

If the German fighters did not rise to attack American bombers, then P-51 units would conduct "fighter sweeps" of enemy airfields to destroy them on the ground. Many young American pilots became aces in the process. Major George E. Preddy, Jr. (1919–1944), flying with the 352nd Fighter Group of the Eighth Air Force, became the highest scoring P-51 ace of the war with 26.83 aerial and five ground victories and the first pilot to destroy six aircraft in one mission.

The P-51 was a war-winning weapon for the United States and it was the product of an international aeronautical heritage that dated back to the 1930s. North American Aviation, an outgrowth of General Aviation Corporation, became a major force under the leadership of "Dutch" Kindelberger, who previously worked at Douglas. Kindelberger moved the company from Maryland to Los Angeles Municipal Airport in southern California in 1935. The company soon became known for its NA-16 two-seat high-performance training airplane that sold steadily in the United States, Europe, and Latin America. Those contracts positioned North American for more opportunities in the world military aviation market.

As military and industrial mobilization reached unprecedented levels in Europe in 1940, Kindelberger found more opportunities for the small company when the British and French came to the United States shopping for military aircraft. Kindelberger rejected a plan to manufacture the obsolete Curtiss P-40 fighter in favor of producing an all-new design. He had the support of the army, specifically the head of procurement, Col. Oliver P. Echols (1892–1954), who facilitated the official arrangement between North American and the Europeans, provided up-to-date research material, and encouraged the company's purchase of advanced wind tunnel data on fuselage design from Curtiss. North American and the Europeans signed the contract for 400 fighter aircraft, designated the NA-73, in May 1940. The first airplane left the North American factory 102 days later.

Under the leadership of North American's new chief engineer, Austrian-born Edgar Schmued (1899–1985), the NA-73 design team incorporated several new innovations. They placed the air intake and the engine radiator on the underside of the fuselage, which made the monoplane even more streamlined and made the internal flow of air so efficient that it actually provided a small measure of forward thrust to the airplane. The Mustang wing featured an airfoil profile that generated the lowest amount of drag up to that time. It was ideally called a "laminar flow" airfoil where the air moved smoothly over the top of the wing in evenly compressed layers while producing maximum lift and minimum drag. NACA researcher Eastman N. Jacobs pioneered the study of laminar flow airfoils in the variable-density wind tunnel at Langley before the war and the North American design team put the data to good use. Finally, the use of straight wingtips and existing components from North American's AT-6 trainer made the P-51 easy to manufacture in large numbers.

The original NA-73 relied upon the same engine used by the Curtiss P-40, the Allison V-12, which provided poor high-altitude performance. At the insistence of the Royal Air Force, North American installed a 1,650-horsepower Rolls-Royce Merlin engine with a two-stage supercharger. The addition made the Mustang the fastest and highest-flying piston fighter of the war at speeds approaching 450 miles per hour and at altitudes up to 40,000 feet. Unable to keep up with production demands, Rolls-Royce licensed the Packard Motor Car Company and Continental Aircraft to produce more than 58,000 Merlin engines in the United States.

The introduction of the Mustang renewed the combined bomber offensive against strategic targets in Germany. "Big Week" of February

1944 witnessed more than 1,000 American bombers acting in concert with the RAF, which initiated a two-month assault on German industry and the Luftwaffe that resulted in the loss of 800 fighters and, more importantly, experienced pilots. Quickly, the Luftwaffe lost its previously unchallenged control of the air over the European continent.

The achievement of Allied air superiority over Europe facilitated the Normandy invasion of June 1944. The Americans and the British successfully airlifted waves of airborne troops into France under the cover of darkness in preparation for the main invasion on D-Day. Their primary mode of transportation was the Douglas C-47 Skytrain, the militarized version of the revolutionary Douglas DC-3 airliner of the 1930s, which could carry an entire detachment of paratroopers or tow a glider full of soldiers and heavy equipment. With their timely arrival, units of the American 82nd and 101st airborne divisions captured crucial objectives that ensured the success of the invasion.

Tactical air operations ensured that the Allied ground advance would continue. Building upon valuable experience gained in North Africa, Sicily, and Italy during the period 1942–1943, the American Ninth Air Force targeted German communications and transportation networks on and near the battlefield. The premier American fighter-bomber was the Republic P-47 Thunderbolt. Alexander Kartveli (1896–1974) originally designed the P-47 as a high-altitude fighter with a 2,000-horsepower Pratt & Whitney radial engine and an elaborate turbosupercharger system. It proved to be an able adversary for German fighters, but its short range, high speed, rugged construction, and heavy armament consisting of eight machine guns, 2,000 pounds of bombs, or ten rockets made it an ideal ground-attack fighter-bomber. From D-Day until the end of the war in Europe in May 1945, Thunderbolt squadrons roamed the European countryside destroying everything they could find, including railway cars and locomotives, armored vehicles, truck convoys, supply depots, and troop concentrations.

Even though they were not allowed to participate in combat operations, women aviators contributed to the air war. Under the leadership of the primary advocates for women military pilots, Nancy Harkness Love (1914–1976) and Jacqueline Cochran (1906–1980), female pilots served in different organizations until being merged into the civilian Women's Air Force Service Pilots (WASPs) in 1943. The WASPs provided basic flight instruction, performed flight tests of new aircraft, towed targets for aerial gunnery training, and delivered newly manufactured aircraft to combat zones so that more male pilots would be available for combat operations. Overall, more than a thousand women served in the WASPs and thirty-eight lost their lives in the line of duty.

AIR WAR IN THE PACIFIC

The victory of American carrier forces at Coral Sea and Midway effectively stopped Japanese expansion into the Pacific and allowed the United States to initiate a dual offensive. The army, honoring the promise made by Army Gen. Douglas MacArthur (1880–1964) to return to the Philippines, would advance through the western Pacific via New Guinea. The navy and marines, led by Adm. Nimitz, would secure the island groups in the Central Pacific with the primary goal of establishing air bases to bomb Japan in preparation for a full-scale invasion. Both campaigns would be examples of unprecedented coordination between land, sea, and air units that would bring the full brunt of American military and industrial power to bear on Japan.

As MacArthur's forces pushed through New Guinea to the Philippines, they benefited from the protection and aggressiveness of the Fifth Air Force under the innovative leadership of Maj. Gen. George C. Kenney (1889–1977). Kenney used new types of weapons and techniques in low-level hit-and-run attacks against Japanese air, land, and sea forces. He ordered the modification of the Fifth's Douglas A-20 Havoc and B-25 Mitchell twin-engine medium bombers with extra machine guns and cannon, protective armor, long-range fuel tanks, and racks for parafrags. The latter were a new type of fragmentation bomb equipped with a parachute that exploded above ground where it would do the most damage and gave the bomber enough time to speed away unharmed. At the Battle of the Bismarck Sea in March 1943, the heavily armed aircraft decimated a Japanese troop convoy with skip bombing, the technique of dropping a bomb at low-level over water to make it "skip" across to its target in much the same way a child would skip a rock across a pond.

The fighter sweep was an integral part of Kenney's tactics. The Lockheed P-38 Lightning was a twin-boom monoplane with a central nacelle that housed the cockpit and armament of four machine guns and one cannon. Each boom housed a high-output Allison V-1710 twelve-cylinder vee engine and an elaborate turbosupercharging system. Overall, the Lightning was a futuristic high-performance fighter capable of speeds approaching 400 miles per hour at altitudes up to 30,000 feet. In the hands of the Fifth's fighter pilots, the Lightning outclassed the Zero and other Japanese designs with its speed and heavy armament. P-38 pilot Maj. Richard I. Bong (1920–1945) became America's highest-scoring fighter ace of World War II with forty victories while flying with the 475th Fighter Group in the southwest Pacific.

After modest beginnings in the 1920s, American naval aviation, specifically the role of the aircraft carrier, became crucial in winning the war in the Central Pacific campaign. The island groups in the Central Pacific were well beyond the range of land-based aircraft, so the importance of the aircraft carrier reached an entirely new level. American industry produced fourteen of the new *Essex*-class and *Independence*-class carriers, capable of carrying 100 and 50 aircraft respectively, and thirty-five smaller escort carriers in 1943 alone. American industry also produced a new generation of American carrier aircraft that were available in great numbers and were superior to Japanese technology. The Grumman F6F Hellcat fighter combined speed, agility, firepower, range, and survivability to achieve approximately 75 percent of the navy's air-to-air victories in the Pacific. The widespread adoption of radar throughout the fleet gave all-weather capability and provided efficient control of air resources during combat operations.

The American aircraft carrier's role in the Pacific war culminated during the Battle of the Philippine Sea, the largest carrier battle in history, in June 1944. The American fleet, under the command of Vice Adm. Raymond Spruance (1886–1969), clashed with the Japanese fleet, but the

Floating airfields: a Grumman F6F Hellcat prepares to take off and attack Japanese targets in the South Pacific. U.S. Navy via National Air and Space Museum, Smithsonian Institution (SI 85-7306).

battle was one-sided. Navy pilots called the battle the Marianas Turkey Shoot because they destroyed three Japanese carriers and more than 400 aircraft at the cost of fifty aircraft. More importantly, the victory led to the capture of the Marianas Islands, which brought army strategic bombers within range of Japan.

The primary objective of the combined army-navy "island hopping" campaign was to get close enough to conduct a mass invasion of Japan. A significant component of the preinvasion strategy was to significantly weaken the Japanese ability to fight and continue the war by attacking industrial and population centers. The key weapon in the strategic bombing campaign against Japan was the Boeing B-29 Superfortress. The Superfortress was the most advanced airplane in the world and a technological marvel. The bomber had its origins in a 1940 air corps call for proposals for a very-long-range bomber with a pressurized cabin and tricycle landing gear. Boeing's submission was a model of streamline design with countersunk rivets covering its tubular fuselage and long and thin cantilever monoplane wings, high-output eighteen-cylinder 2,200-horsepower Wright R-3350 radial engines, efficient flaps for takeoff and landing, and a highly advanced remote control system for the defensive armament. The new bomber could carry 16,000 pounds of bombs and cruise 235 miles per hour at altitudes up to 30,000 feet. The prototype first flew in September 1942. By all accounts, the B-29 was the epitome of the modern airplane that first emerged with the introduction of the Douglas DC-3 in the mid-1930s.

Overall, the B-29 program would cost over $3 billion ($31 billion in modern currency), exceeding the cost of another high-priority and supersecret American project, the Manhattan District, or the atomic bomb. Despite the cost and government priority, the B-29 faced development problems, primarily with its radial engines. The R-3350s would overheat and catch fire. One engine fire cost Boeing the life of its chief test pilot when it brought the lumbering bomber down in Seattle. Continued changes to the airframe in preparation for combat operations threatened the overall production schedule. Factories in Washington, Kansas, Nebraska, and the integrated plant operated by Bell Aircraft near Atlanta, produced thousands of B-29s for the war effort. During a dramatic six-week period in March and April 1944, the army had to fight the "Battle of Kansas" to get the first operational B-29s ready for service.

B-29 operations first began in India and China in June 1944. The lack of a direct supply line and inadequate airfields led to a shift to the islands of Saipan, Tinian, and Guam in the Marianas after their bloody capture by American forces in the fall. Unlike the other operational air forces, the bomb groups of the Twentieth Air Force in the Marianas were a purely strategic force whose leaders reported directly to the chief of the army air forces, Gen. Arnold.

As soon as the campaign began, it was clear that the operational environment over Japan would deter the success of daylight, high-altitude, precision bombing. The commander of the B-29 groups, Gen. Haywood S. Hansell (1903–1988) intended to follow air force doctrine to the letter. His units would attack specific industrial and logistical targets while avoiding population centers and other areas not directly connected to the Japanese war effort. From the Marianas, it was a 3,000-mile round trip to Japan over water. It was not until after the bloody capture of Iwo Jima in February 1945 that bomber crews had a safe haven at the midway point and a base for their P-51 Mustang escort fighters. The weather of the northwestern Pacific obscured targets. The effects of the jet stream at high altitude, the strong and steady current of winds found at high altitude, catapulted B-29s quickly over their targets or made them motionless if they faced it head on. Most of the architecture in Japan was light wood structures that high-explosive bombs rarely affected. Interspersed in civilian housing, seen as off-limits in American strategic bombing theory, were industries crucial to the Japanese war effort. Arnold relieved Hansell in early 1945 due to his failure to achieve results with daylight, high-altitude, precision bombing.

Hansell's replacement, Gen. Curtis E. LeMay (1906–1990), a veteran of bomber command in Europe and China, adapted tactics to fit the conditions the B-29s faced over Japan. He switched the focus from daylight, high-altitude, precision attacks to low-level, nighttime, incendiary raids with the B-29s stripped of most of their armament to carry more bombs and their undersides painted black. By simply burning the mostly wood cities to the ground, the Twentieth would destroy Japanese industry and kill or drive away its workers in the process. The Tokyo raid in March 1945 alone killed upwards of 100,000 people and destroyed one-fourth of the city in a single twenty-four hour period. The addition of daylight raids on selected targets and the aerial mining of the shipping lanes surrounding Japan effectively left the island nation isolated and in economic, political, military, and social shambles by the end of July 1945.

Starting in Europe with the British and American raids on the cities of Hamburg and Dresden, where over 135,000 German civilians perished in two days, the army air forces consistently increased its pattern of attacking Japan through area bombing. Area bombing consisted of a bomb group delivering its deadly cargo in the general vicinity of a target without taking into consideration the location of civilians. Oftentimes, the civilians were the target. Area bombing was unrestricted and without moral restraint as the B-29s of the Twentieth Air Force waged total aerial war on all of Japan.

The addition of a new type of weapon, the atomic bomb, and its unprecedented ability to destroy an entire city, ensured that the United States would use any weapons at its disposal to defeat Japan. In a program equal in size, scope, and secrecy to the B-29 development program, the United States created the world's first atomic bombs. Centered on the government laboratory at Los Alamos, New Mexico, the American program culminated in a successful test of a plutonium bomb in July 1945, and with it the opportunity to usher in the atomic age by attacking Japan and effecting a rapid end to the war.

The army air forces created the 509th Composite Group based at Tinian as the world's first atomic bombing force. Their commander, Col. Paul W. Tibbetts, Jr. (1915–), led their intensive training in the United States and in the Pacific for one specific mission: to deliver an atomic bomb to a target in Japan. On August 6, 1945, Tibbetts and his crew took off in their B-29, named *Enola Gay* for Tibbetts's mother, and dropped their single bomb, called Little Boy, on the city of Hiroshima. That one uranium bomb killed approximately 60,000 people instantly, and another 60,000 died later from radiation sickness and related injuries. Additionally, the bomb left less than 20 percent of the city's buildings standing. Three days later, on August 9, 1945, another 509th B-29, named *Bock's Car* after a previous pilot, with Maj. Charles Sweeney (1919–2004) at the controls, left Tinian and attacked the city of Nagasaki. The second bomb, called Fat Man for its large, rotund shape and made with plutonium, killed approximately 35,000 people instantly, and another 40,000 died from sickness and injuries. Facing the open declaration of war by the Soviets and the threat of continued incendiary and atomic bombings, the Japanese government formally surrendered to American military leaders aboard the battleship *Missouri* anchored in Tokyo Bay on September 2, 1945.

The atomic attacks on Hiroshima and Nagasaki and their effect on ending the war in the Pacific have been shrouded in controversy. Military and political leaders, as well as generations of veterans since, argued that using the atomic bombs against Japan shortened the war and prevented a large-scale invasion that would have resulted in the loss of hundreds of thousands of American lives. They saw the mass aerial suicide attacks by the Japanese Kamikaze units, which sank thirty-six ships and incurred 5,000 casualties as an indication of Japan's resolve to continue the war. Critics of the decision to use the bomb recognized the awesome power and responsibility created by the availability of "the bomb." The bombs are a stirring example of how technology, normally celebrated to a high degree before the war as a positive force in history, could also symbolize the utter destruction of humankind itself. It is hard to disassociate from two triumphs of modern technology—the modern airplane and atomic energy—the chilling fact that over the course of three days, two B-29s with two atom bombs had killed almost 200,000 people in August 1945.

5

The Second Aeronautical Revolution, 1930–Present

◆

In July 1942, test pilot Fritz Wendel (1916–1975) took to the skies over Nazi Germany in a Messerschmitt Me 262, a revolutionary new type of airplane. With its swept wings and two gas turbine engines, the Me 262 was the world's first practical jet airplane. Its appearance during World War II ushered in a second Aeronautical Revolution and the next great age in aviation history, the jet age. The invention of the turbojet engine and the requisite engineering to make it and the jet aircraft that followed viable would be equal to the achievement of the Wright brothers. From 1930 and on throughout the twentieth century, the airplane became increasingly sophisticated as the aeronautical community refined it to fly higher, faster, and farther than ever before.

BIRTH OF THE JET AIRPLANE

The idea of jet propulsion was not new. Precedents for gas turbine engines dated back to the late 1820s with high-speed water and steam turbines for industrial and marine use in Europe. During the Great War, French engineer Auguste Rateau (1863–1930) created a turbosupercharger for aircraft engines, a device that compressed air before it entered the cylinders to increase engine power and performance at high altitudes. In the United States,

The Messerschmitt Me 262 represented the dramatic changes in high-speed aeronautical technology. National Air and Space Museum, Smithsonian Institution (SI 76-13231).

the army sponsored a turbosupercharger program led by Sanford A. Moss (1872–1946) of the General Electric Company (GE) at Lynn, Massachusetts. Moss's pioneering work in engineering high-temperature alloys and thermodynamic design would become important during the World War II era when the majority of Allied combat aircraft such as the B-17 and the P-38 employed complex turbosupercharger systems.

Prototypical jet-powered aircraft also began to take shape in the early twentieth century. These early designs replaced the propeller with an arrangement called a thermojet, where an air inlet, radial piston engine, compressor, combustion chamber, and exhaust outlet generated thrust for power. Henri M. Coanda (1886–1972), a Rumanian engineer, first exhibited a biplane in 1912 with an elaborate circular duct at the front to demonstrate the principle, but the craft never flew. In Italy, Secondo Campini (1904–1980) designed a similar system that ran the entire length of a specially designed airplane, the Caproni-Campini CC.2. The new airplane flew at speeds up to 200 miles per hour in August 1940, which was much slower than any propeller-driven aircraft of the period.

Despite those early steps toward making a jet engine and an airplane to go with it, it would be two separate individuals, one in isolation and the other at a leading aircraft manufacturer, to realize the dream of jet-powered flight. A Royal Air Force officer, Frank Whittle (1907–1996), sought an alternative to the piston-engine and propeller combination. Unlike many of his fellow officers, he did not come from a privileged background and rose up through the ranks. As part of his training at the RAF's technical college from 1926 to 1929, Whittle authored a paper entitled "Future Developments in Aircraft Design." He came to the conclusion that according to Newton's third law of physics, a gas turbine could produce jet propulsion to power an airplane at higher speeds than previously thought possible. He patented his idea later in 1930. The British Air Ministry ignored Whittle's work and his patent went unnoticed by the international aeronautical community. As a result, he allowed the patent to lapse in a moment of frustration. Whittle went on to earn an advanced degree in mechanical sciences at Cambridge. He founded Power Jets, Ltd., in March 1936 with a small group of investors and the encouragement of his Cambridge professors. The sole purpose of the company was to design, build, and test a jet engine. The first complete engine, the W.U., or Whittle Unit, ran on a test stand in April 1937. With a renewed interest, the Air Ministry contracted Power Jets to build a flying engine while Gloster Aircraft received another contract to build a jet-propelled airplane.

While Whittle studied at Cambridge, Hans Pabst von Ohain (1911–1998) earned his doctoral degree in aerodynamics and physics from the leading German technical school, Göttingen University, in 1935. Independently, he reached the same conclusions as Whittle and patented his own aeronautical gas turbine engine. Using private funds, he and a friend constructed a promising, but small, working model. With a letter of introduction from his mentor at Göttingen, von Ohain met with German aviation industrialist Ernst Heinkel (1888–1958), who quickly gave the young physicist a job developing the company's, and Germany's, first jet engine in March 1936. Heinkel had a widely known obsession with building high-speed aircraft.

The collaboration between von Ohain and Heinkel moved quickly. Initially using hydrogen as a fuel, von Ohain moved to gasoline to make the engine practical for flight. Von Ohain tested his first working engine in March 1937. Heinkel engineers designed a purpose-built airframe for von Ohain's Heinkel HeS 3B engine. The Heinkel He 178 flew in August

1939, making it the world's first gas turbine-powered, jet-propelled airplane in history to fly.

Due to development difficulties centered on the hydrogen fuel system and institutional resistance, Whittle's engine, the W.I, would not fly in the Gloster E.28/39 until May 1941, well after the outbreak of World War II. The British went on to refine the Whittle design. The twin-engine, single-seat Gloster Meteor jet fighter became the first and only turbojet-powered airplane to serve with the Allies during the war when it entered operational service in August 1944. Capable of 480 miles per hour, the Meteor influenced an entire generation of British jet aircraft.

Despite not knowing about each other's work, both von Ohain and Whittle designed turbojet engines with a centrifugal-flow compressor. Air entered through the front of the engine where a rotating impeller moved it outward into compression chambers. From there, the fuel/air mixture traveled to combustion chambers for ignition. The resultant thrust drove the turbine as it exited the engine through the exhaust outlet. The design of a centrifugal-flow engine was complicated since the air flowed through twists and turns in a roundabout fashion, the shape was bulky, and it was not fuel efficient. Both the von Ohain and the Whittle engines delivered similar performance. The HeS 3B engine generated 838 pounds of thrust and propelled the He 178 at speeds up to 360 miles per hour. The Whittle W.IX produced 860 pounds of thrust, which gave the E.28/39 a top speed of 338 miles per hour.

The success of the He 178 and von Ohain's engine led the German government to encourage the development of jet-propelled aircraft during World War II. The most successful design, the Messerschmitt Me 262, was the world's first practical jet airplane. The revolutionary design and performance of the Me 262 was a stunning example of what jet aircraft could and would do over the next fifty years of flight.

The original Me 262 design of 1938 featured straight wings, jet engines mounted in the center of the wings, and conventional landing gear with a tail wheel at the rear of the fuselage. Continuous changes in the airframe design delayed the program and were still taking place at the time of the first flight on pure jet power in July 1942. The final configuration was a distinctive and sleek single-seat, low-wing, all-metal monoplane powered by two jet engines housed in underslung nacelles and carrying a powerful armament of four heavy cannon. The high-speed swept wings measured 40 feet 11 inches and included slats, or movable airfoils, along the leading edge for increased maneuverability at low speeds. The triangular fuselage was 39 feet 9 inches long. The tricycle landing gear, which aided controllability on the ground and put the aircraft in an instant flight attitude, put the aircraft at a height of 12 feet 7 inches.

Two Junkers Jumo 004B turbojet engines powered the Me 262. Junkers Motoren began work on the new design under the direction of Dr. Anselm Franz (1900–1994), the head of the company's supercharger group, in 1939. Franz's engine utilized an axial-flow compressor, an alternating series of rotating and stationary blades where the overall flow of air essentially moved along the axis of the engine as it was compressed and then ignited. At the end of the engine was the world's first afterburner, an extension of the exhaust section that ignited leftover fuel to produce even more thrust. The Jumo 004B design became the standard configuration for all modern jet engines and was the first to be produced in large numbers.

Problems plagued the Jumo 004 development program due to the nonavailability of high strength and high temperature alloys and changes needed for volume production, which delayed the operational introduction of the Me 262. Nevertheless, with a Junkers Jumo 004B turbojet engine mounted on each wing generating 1,980 pounds of thrust, the Me 262 could fly at speeds up to 540 miles per hour. That was more than 100 miles per hour faster than any contemporary propeller-driven, piston-engine airplane.

The Luftwaffe, facing the onslaught of hundreds of Allied fighters and bombers attacking around-the-clock, knew that it could not produce enough conventional piston-engine propeller-driven aircraft to defend Nazi Germany. Quickly, it became apparent that the Me 262, and similar jet designs, would be the only type of aircraft that could potentially maintain air superiority over Europe. The leading Luftwaffe fighter aces, including their leader, Maj. Gen. Adolf Galland (1912–1996), believed they could resist the Allies in the sky with the fighter version of the Me 262, called the Schwalbe, or Swallow. At that critical moment, Hitler's misplaced enthusiasm for a fighter-bomber version of the Me 262, the Sturmvogel (Stormy Petrel), threatened the introduction of the air defense fighter. It was only after considerable persuasion on the part of the pilots that Hitler allowed a dedicated focus on the Schwalbe.

Me 262 squadrons began combat operations in July 1944, but they did not have the effect Galland and his fellow pilots envisioned. Nazi Germany suffered from disrupted communications, the overall lack of supplies, fuel, and qualified pilots. Innovative tactics ensured that the Allies would retain air superiority over Europe. Rather than trying to out-fly the Me 262, P-51 pilots used the fighter sweep to attack them at their most vulnerable moments in the air, when they were landing or taking off. The almost absolute Allied dominance of the air offset the Me 262's performance advantage. The revolutionary fighter would not win the war for the Nazis.

AMERICA AND THE JET

The United States, lagging behind Germany and Great Britain in jet aircraft development, urgently needed to get a combat-ready jet into the air during World War II. These first major American jet aircraft designs relied upon the centrifugal-flow engines of British origin. Early in 1941, GE and army representatives working with the Royal Air Force in England learned of the British turbojet program. General Arnold saw the flight of the Gloster E.28/39 and convinced the British government to assist the United States in jumpstarting its gas turbine program. The British government sent an early Whittle engine and drawings of the latest design to General Electric at Lynn in October 1941. GE's expertise in turbine and turbo-supercharger technology made it the obvious choice to develop the American military's first jet engine, the I-A. The new engine produced 1,250 pounds of thrust.

Two GE I-A engines powered America's first jet airplane, the Bell XP-59A Airacomet. A single-seat, all-metal monoplane, the Airacomet exhibited performance equal to the best piston-engine propeller-driven fighters during its first test flights beginning in October 1942. GE introduced the improved I-16 engine, which generated 1,600 pounds of thrust and enabled operational P-59 aircraft to fly at speeds up to 450 miles per hour in July 1943.

While the Airacomet proved to be a difficult start, its successor, the Lockheed P-80 Shooting Star proved to be not only the first practical American jet airplane, but one of the most significant American aeronautical designs in history. The U.S. government awarded a contract for a fighter powered by the British de Havilland Goblin H-1B turbojet engine to Lockheed Aircraft Corporation as an improvement over the Airacomet. Clarence L. "Kelly" Johnson (1910–1990), chief research engineer for Lockheed, and a group of talented engineers and technicians designed and constructed the prototype, designated the XP-80, in a remarkable 143 days with its first flight in January 1944.

Named *Lulu-Belle* after a character in the widely famous Al Capp comic strip *Li'l Abner* (1934–1977), the new dark-green fighter featured a tapered, elliptical nose, thin straight wings, and smooth, faired air intake ducts. *Lulu-Belle* featured an innovative removable tail that permitted easy access to the de Havilland Goblin H-1B centrifugal-flow turbine engine capable of producing 2,240 pounds of thrust. The Goblin was never available in sufficient numbers and not powerful enough for subsequent variants of the XP-80. Nevertheless, *Lulu-Belle* was the first aircraft in the United States to exceed 500 miles per hour in level flight. Flying the XP-80 gave the American military and the industry valuable experience in designing, manufacturing, and operating military jet fighter aircraft.

By the end of 1945, the XP-80 was a resounding success for Johnson and his team of designers and engineers. They had been working independently within Lockheed and created a distinctive management style that focused on every part of military aircraft design and manufacturing from the drawing board to flight test under the highest levels of secrecy and security. They were officially known as the Advanced Development Projects within Lockheed, but they collectively called themselves the Skunk Works, an adaptation of the name for the ultra-secret, high-security "Skonk Works" moonshine liquor still featured in *Li'l Abner*. The streamlined organization, which emphasized simplicity, military-industrial cooperation, and responsible management, became a model for success in a rapidly expanding, and hard to control, aerospace industry. At the core of the success of the Skunk Works was Johnson, who wanted a direct relationship between design engineer and mechanic and manufacturing.

Johnson was one of the foremost American aircraft designers of the twentieth century. Twice winner of the Collier Trophy and many other prestigious aerospace awards, his remarkable career at the Lockheed Aircraft Corporation spanned from the Aeronautical Revolution of the 1930s to the Cold War, and the edges of space. Throughout that long period, his design work was consistently innovative and futuristic. At the University of Michigan, Johnson studied aeronautical engineering and joined Lockheed in 1933 at the age of 23. His work in the GALCIT wind tunnel led to the success of the twin-tail Electra airliner. Johnson codesigned the P-38 Lightning, one of the most versatile and revolutionary fighters of World War II. After the creation of the Skunk Works and *Lulu-Belle*, Johnson remained to be a major player within the aviation industry for the next thirty years.

The production Lockheed P-80 Shooting Star became America's first operational jet fighter and, overall, the first practical military jet airplane. It was larger and more powerful than the original *Lulu-Belle* with the new GE I-40 turbojet capable of generating 3,750 pounds of thrust. The Shooting Star and the I-40 were the first jet system deemed suitable for quantity production by the army. Unfortunately, they were not ready for operations in World War II, but Shooting Stars did serve as the primary fighter for the army air forces and its successor, the newly created U.S. Air Force, in the late 1940s. The P-80 proved to be a versatile and stable design adaptable to a variety of combat and peacetime missions. The elongated nose design allowed the concentration of six machine guns in the nose of the aircraft. Cameras replaced guns for photoreconnaissance missions. The hinged nose section allowed access to the cameras. Easy to fly, it was an ideal aircraft in which to train thousands of military pilots transitioning from propeller-driven aircraft to jets.

The first jet engine of all-American design appeared in 1944. The navy worked directly with the newly formed Aviation Gas Turbine Division of the Westinghouse Electric Corporation of Philadelphia to create the 19A turbojet, which featured a rotary compressor driven by a turbine wheel with axial rather than centrifugal compressors. The small engine developed 1,200 pounds of thrust and first flew as booster for a navy fighter in January 1944. Two improved 19B turbojets, each capable of generating 1,365 pounds of thrust, powered the FH-1 Phantom, the first jet designed for naval operations and designed by the McDonnell Company in St. Louis, in January 1945. The Phantom made the first all jet-powered takeoff and landing from the carrier USS *Roosevelt* in July 1946.

The first American jets were taking to the air as World War II ended. With the fall of the Nazis, the victorious Allies—the United States, Great Britain, and the Soviet Union—were poised to reap the benefits of advanced German aeronautical research. The army air forces carried out Operation LUSTY (*Luftwaffe Secret Technology*) to acquire advanced German aircraft and bring them back to the United States. The pilots participating in the operation, known as Watson's Whizzers after their commander, Col. Harold E. Watson (1911–1994), were proud to say that they were America's first unofficial fighter jet squadron. To indicate that, they removed the propellers from their standard army air forces shirt collar insignia and wore a patch on their leather flight jackets that featured the famous Walt Disney character Donald Duck holding on to a Junkers Jumo 004B engine as it circled the earth.

Operation LUSTY recovered many examples of advanced aeronautical design, including the Me 262, another jet fighter called the Heinkel He 162 A-2 Spatz (Sparrow), and the world's first operational jet bomber, the Arado Ar 234 Blitz (Lightning). Once in the United States, these aircraft went to the army air forces research and development facility at Wright Field near Dayton, Ohio. These aircraft, and the aerodynamic research that made them so revolutionary, influenced a new generation of American military jet aircraft. Both the North American Aviation F-86 Sabre fighter and the larger Boeing B-47 Stratojet that appeared in 1947 featured swept wings, wing slats, and adjustable stabilizers. The merging of the

German tradition with a rapidly maturing American tradition of high-speed aeronautical design resulted in a second Aeronautical Revolution.

HIGH-SPEED FLIGHT AND THE RESEARCH AIRPLANE

The high speeds made possible by the new jet technology highlighted problems with aircraft design already recognized by the aeronautical community. Many in the 1930s and 1940s believed in the existence of an invisible barrier in the sky that prevented aircraft from flying faster than the speed of sound, or Mach 1, the ratio of the speed of sound to the speed of an object, which was approximately 700 miles per hour. Government researchers at McCook Field led by Frank W. Caldwell (1889–1974) and at the U.S. National Bureau of Standards under the direction of Lyman J. Briggs (1874–1963) and Hugh L. Dryden (1898–1965) identified the problem first on aircraft propellers in the 1920s. As the tips of a whirling propeller approached the speed of sound, they suffered from a marked decrease in efficiency—a loss of lift caused by increased drag—called "compressibility burble." Propellers simply could not turn any faster. The aeronautical community saw it as a minor problem because, at the time, aircraft could fly no faster than 200 miles per hour. As aircraft speeds increased toward 400 miles per hour in the early 1930s, researchers investigated the problem of compressibility at supersonic speeds to determine its effect on aircraft performance. As an aircraft approached the speed of sound, the air in front of it compressed to create shock waves that spread outward across the airplane. Those shock waves disrupted the airflow over the wing, which caused compressibility and rendered the aircraft uncontrollable.

Europe quickly became a center for high-speed aerodynamics. Jakob Ackeret (1898–1981) operated the first wind tunnel capable of generating Mach 2 in Zurich, Switzerland. In Germany, the universities at Göttingen and Aachen provided a strong academic tradition of theoretical aerodynamics and the Nazi government's aeronautical research facility, the Deutsche Versuchsanstalt für Lufthart, cut a clear path of advanced development that quickly outpaced other countries. This work included not only aircraft research, but also innovative work in rockets and missiles.

The international high-speed aerodynamics community met to discuss their individual work in 1935 at the Volta Conference held in Rome. One of the papers presented a revolutionary answer to the problem of high-speed aerodynamics and the sound barrier. German aerodynamicist Adolf Busemann (1901–1986) argued that if aircraft designers swept the wing back from the fuselage, it would offset the increase in drag beyond speeds of Mach 1. The Volta Conference proved to be a turning point in high-speed aerodynamics research. Nazi Germany emphasized fundamental research that created a new generation of advanced high-speed aircraft. The United States, however, concentrated on perfecting and refining traditional propeller-driven aircraft with straight wings.

American propeller-driven aircraft did suffer from aerodynamic problems caused by high-speed flight during World War II. Flight testing of the P-38 Lightning revealed compressibility problems that resulted in the death of a test pilot in November 1941. As the Lightning dove from 30,000 feet, shock waves formed over the wings and hit the tail causing violent vibration, which caused the airplane to plummet into a vertical, and unrecoverable, dive. Extensive wind tunnel testing by the NACA at its new Ames Laboratory in 1944 led to the modification of the airplane with fillets, fairings that smooth the flow of air between the fuselage and wings, and dive recovery flaps to offset compressibility.

Almost ten years after the Volta Conference, American researcher Robert T. Jones (1910–1999) independently discovered the swept wing at NACA Langley. Jones was one of the most respected aerodynamicists at Langley and his five-page 1945 report provided a standard definition of the aerodynamics of a swept wing. The report appeared at the same time high-speed aerodynamic information from Nazi Germany was reaching the United States.

With the end of World War II in Europe in May 1945, a team of American scientists and engineers searching for German aeronautical secrets uncovered a cache of aerodynamic data on swept wings at a laboratory at Braunschweig, Germany. One team member was a young aeronautical engineer employed by Boeing Aircraft, George Schrairer, who had been working on a new jet-propelled bomber design. Schrairer informed his colleague Ben Cohn back in Seattle of the new data. The new Boeing B-47 Stratojet bomber, originally designed with a conventional straight wing, now had swept wings due to the German data. There was an American contribution to the B-47's design. The four engine pods hung from the wing on pylons were the result of extensive work in the company wind tunnel. As a result, the revolutionary 35-degree wing sweep and the engine pods became signature characteristics of all multiengine jet aircraft that would follow.

As the German and American traditions merged into the B-47, the American aeronautical community realized that there were still many questions to be answered regarding high-speed flight. It became apparent to NACA researcher John Stack (1906–1972) as early as 1933 that building a full-scale "research airplane" would be a valuable way to investigate unknown flight regimes. Working with Stack, army engineering officer Ezra Kotcher (1903–1990) proposed a full-scale flight program to explore transonic and supersonic flight with the specific goal of achieving Mach 1 in 1944.

Bell Aircraft Corporation of Buffalo, New York, received the contract to design and build the XS-1 (*Experimental Sonic 1*) in early 1945. With twenty years of aerodynamic data provided by the NACA, Bell engineers shaped the fuselage after a .50-caliber machine gun bullet, an object known to fly at supersonic speeds. The wings were straight, thin, and measured only 28 feet. The XS-1 featured a new adjustable horizontal stabilizer, or "flying tail," that could be moved up and down for better control at transonic speeds, which became a standard component for military aircraft. Rather than a conventional propeller and piston engine or a jet engine, the new airplane relied upon a 6,000-pound-thrust Reaction Motors XLR-11 rocket engine to push it through the sound barrier.

On October 14, 1947, a B-29 bomber took off from Muroc Dry Lake in the Mojave Desert in California. Hanging in the bomb bay were the XS-1 and its pilot, Capt. Charles E. "Chuck" Yeager (1923–). A Mustang

Chuck Yeager and the Bell XS-1 broke through the sound barrier in October 1947. National Air and Space Museum, Smithsonian Institution (SI 97-17485).

pilot during the war, Yeager named his research airplane *Glamorous Glennis* after his wife. After being dropped from the B-29 and igniting the rocket engine, Yeager became the first pilot to enter safely the unknown realm of the sound barrier when the XS-1 reached Mach 1.06 (700 miles per hour) at an altitude of 43,000 feet. Yeager, Lawrence D. Bell (1894–1956), and John Stack received the 1947 Collier Trophy for the stunning example of the partnership among the military, the aviation industry, and the government researchers of the NACA in achieving supersonic flight. Yeager would go on to be immortalized in Tom Wolfe's book (1979) and the subsequent film (1983) *The Right Stuff* as the heroic military test pilot.

After the flight of the XS-1, the aviation industry, the NACA, and the military services operated a new family of research airplanes at Edwards Air Force Base in the high desert of California that sought to answer questions about high-speed flight. The swept wing Douglas D-558-II Skyrocket achieved Mach 2 in 1953. The Douglas X-3 Stiletto, designed for extended Mach 2 flight, featured a needle nose and short stubby wings. Many of the research airplanes investigated innovations pioneered by the Germans during the war, including the swept-wing tail-less aircraft (Northrop X-4 Bantam), movable, or variable-sweep, wings (Bell X-5), and the triangular delta wing (Convair XF-92A).

One of the major aerodynamic challenges of the 1950s was bridging the gap between subsonic and supersonic speeds. Much was known about those regimes, but little was known about the regime in between, the transonic regime. The air force's new YF-102 Delta Dagger delta wing fighter designed by Convair of San Diego, California, was unable to reach supersonic speeds during its first flights in 1952. A substantial air force contract and Convair's reputation were at stake. A mild-mannered NACA researcher named Richard Whitcomb (1921–), who had been studying the problem of transonic drag, used a special tunnel developed by Stack at Langley. He realized that the increase in drag as an airplane approached supersonic speeds was not the result of shock waves forming at the nose, but by those forming just behind the wings. Whitcomb asserted that narrowing, or pinching, the fuselage where it met the wings would decrease transonic drag and NACA transonic wind tunnel research supported his idea. The improved YF-102A with its new area rule fuselage achieved supersonic flight during its first flight in 1953. Moreover, the area rule fuselage increased the Delta Dagger's top speed by 25 percent. Whitcomb's revolutionary idea enabled a new generation of military aircraft to achieve supersonic speeds.

FLYING INTO SPACE

The Soviet launch of the Sputnik satellite into the Earth's orbit in 1957 initiated a major change in American aeronautical research and development. The National Air and Space Act of July 1959 converted the NACA into the National Aeronautics and Space Administration (NASA). A primary goal of NASA was to create and then manage America's civilian space program that would compete with the Soviets in putting the first humans in space and ultimately on the moon. The first A in NASA, aeronautics, dealt with improving flight in the atmosphere.

One of NASA's first flight research studies, the X-15 program (1959–1968), investigated hypersonic flight at five or more times the speed of sound at altitudes reaching the fringes of space. Launched from the wing of a Boeing B-52 mother ship, the X-15 was a true "aerospace" plane with performance that went well beyond the capabilities of existing aircraft within and beyond the atmosphere. North American Aviation of Los Angeles, California, had a special challenge in designing the X-15. For propulsion, a Reaction Motors XLR99 rocket engine produced 57,000 pounds of thrust. At hypersonic speeds, the air traveling over an airplane generated enough friction and heat that the outside surface of the airplane reached a temperature of 1,200 degrees

The hypersonic North American X-15 is the world's fastest piloted vehicle. National Air and Space Museum, Smithsonian Institution (SI 89-5928).

Fahrenheit. North American used titanium as the primary structural material and covered it with a new high temperature nickel alloy called Inconel-X. The X-15 relied upon conventional controls in the atmosphere, but used reaction-control jets to maneuver in space.

The long, black research airplane with its distinguishing cruciform tail became the highest flying airplane in history. In one flight, the X-15 flew to 67 miles (354,200 feet) above the earth at a speed of Mach 6.7, or 4,534 miles per hour. At those speeds and altitude, the X-15 pilots, made up of leading military and civilian aviators, had to wear pressure suits and many of them earned the right to wear astronaut's wings. Before he became the first man on the moon in 1969, naval aviator and test pilot Neil A. Armstrong (1930–) reached the edges of space flying the X-15. Overall, the 199 flights of the X-15 program generated important data on high-speed flight and provided valuable lessons for the NASA's space program.

Hypersonic flight as a gateway in and out of the earth's atmosphere was possible. The air force's failed X-20 Dyna-Soar project was an attempt to develop a winged spacecraft that became ensnarled in bureaucratic entanglements. One obstacle to successful reentry was aerodynamic heating. Revolutionary research by NASA researchers H. Julian Allen (1910–1977) and Alfred J. Eggers, Jr. on ballistic missiles revealed that a blunt shape would make reentry possible. NASA developed a series of "lifting bodies" to test unconventional blunt configurations—capable of reentry and then being controlled in the atmosphere. The blunt nose and wing-leading edge of the Space Shuttles that are launched into space and then glide to a landing after reentry, starting with *Columbia* in April 1981, owe their success to the lifting bodies test flown by NASA in the 1960s and 1970s.

The latest NASA research program, called Hyper-X, investigated hypersonic flight with a new type of aircraft engine, the X-43A scramjet, or supersonic combustion ramjet. The previous flights of the X-15, the lifting bodies, and the Space Shuttle relied upon rocket power for hypersonic propulsion. A conventional air-breathing jet engine, which relies upon the mixture of air and atomized fuel for combustion, can only propel aircraft to speeds approaching Mach 4. A scramjet can operate well past Mach 5 because the process of combustion takes place at supersonic speeds. Launched on rocket booster from a B-52 at 40,000 feet, the X-43A flew for the first time in March 2004. During the 11-second flight, the little engine reached Mach 6.8 and demonstrated the first successful operation of a scramjet. Later, in November 2004, a second flight achieved Mach 9.8, the fastest speed ever attained by an air-breathing engine. Much like Whittle and von Ohain's turbojets, the X-43A offered the promise of a new revolution in aviation, that of high-speed global travel.

6

The Jet Airplane as a Military Weapon, 1945–Present

◆

During Operation Desert Storm in early 1991, a small, gray, propeller-driven airplane called Pioneer flew over Faylaka Island in the Persian Gulf off the shores of Kuwait. It was there to assess damage to the island after being bombarded by the battleship USS *Missouri*. Upon seeing the Pioneer, Iraqi troops on the island quickly surrendered. There were no humans aboard Pioneer. It was an uninhabited aerial vehicle (UAV) and for the first time in history human beings had laid down their arms to a machine. Pioneer was but one of many new weapons that emerged during the fifty years since the end of World War II.

THE COLD WAR AND THE ATOMIC MENACE

America emerged from World War II the greatest airpower on earth. The National Security Act of 1947 created a unified Department of Defense responsible for the administration and control of the army, navy, and an independent air force that would primarily be responsible for the use of atomic weapons in future wars. British prime minister Winston Churchill recognized the creation of an "Iron Curtain" across Eastern Europe by the Soviet Union in 1946, which acquired its first atomic bomb in 1949. With both superpowers possessing these new weapons, it became clear that the world would not survive another global conflict. The mutual destruction of both nations, and with them the rest of the world, became a reality in a new "atomic age" that deterred either side from instigating a nuclear conflict. The new war between the United States and the Soviet Union would be a "cold war."

The Strategic Air Command (SAC), under the command of Gen. Curtis E. LeMay from 1948 to 1957, was America's leading deterrent against the Soviet Union. His strategy in the event of a third world war was a massive retaliatory nuclear strike delivered via intercontinental bombers and missiles to destroy the Soviet Union. The frontline bombers for SAC reflected the two heritages of World War II. The Consolidated B-36 Peacemaker, an aerial behemoth with a crew of fifteen and powered by six reciprocating piston engines and four jet engines, plodded along in the air at a little over 200 miles per hour, but could deliver almost 90,000 pounds of nuclear bombs. With a crew of three, the Boeing B-47 Stratojet could deliver 20,000 pounds of nuclear weapons at 500 miles per hour. These aircraft could fly intercontinental missions nonstop due to the introduction of in-flight refueling where aerial tankers replenished the fuel tanks of SAC bombers in midair for truly unlimited global missions. LeMay did not overlook the public relations potential of working with Hollywood when he gave full support and cooperation to the production of the Technicolor epic *Strategic Air Command* (1955), which featured the B-36 and the B-47 as prominently as the star of the film, James "Jimmy" Stewart (1908–1997).

The first major encounter between the Soviet Union and the United States came in 1948 and it tested the new air force. Immediately after World War II, the victorious Allies divided Germany and its capital, Berlin, into eastern and western halves. In disagreement with the Soviets over the future of conquered Germany, the United States, Great Britain, and France created West Germany from their occupation zones. In retaliation, Soviet dictator Joseph Stalin ordered a ground blockade of Berlin, located in the Soviet zone, in preparation for occupying the entire city. With all road, rail, and water connections severed, the Americans and the British staged an around-the-clock airlift of Berlin from June 1948 to May 1949. For almost a year, they kept 2.5 million people fed and warm with 2.3 million tons of supplies during the Berlin Airlift, or Operation Vittles. Operation Little Vittles was part of the larger operation, which involved the dropping of 250,000 pieces of candy by miniature parachutes, made from handkerchiefs and scraps of cloth, to the city's children. The Berlin Airlift offered a peaceful alternative to war, signified the important role of airpower in postwar diplomacy, and gave the new air force a valuable lesson in aerial supply and transport.

As international tensions rose, the U.S. government prepared the civilian population for a nuclear holocaust through an extensive civil defense network that urged citizens to "duck and cover" at the sight of an atomic blast and to build bomb shelters for individual protection. It was not all so serious in the late 1940s and 1950s. America's soldiers had come home and were ready to start their individual lives with new families, homes, and jobs. For the new consumer culture, industrial designers offered new products inspired by the airplane. Automotive stylist Harley Earl (1893–1969) and his colleagues at General Motors, inspired by the Lockheed P-38 Lightning fighter, started a very popular trend that persisted through the 1950s and 1960s when they added twin tail fins to the 1948 Cadillac Fleetwood sedan.

COLD TO HOT: AIR WAR IN KOREA AND VIETNAM

The Cold War escalated into an actual conflict during the Korean War (1950–1953). In the aftermath of World War II, the victorious Allies divided Korea, occupied by the Japanese since the early twentieth century, at the 38th Parallel. Communist North Korea, under the leadership of Kim Il Sung, received support from both the People's Republic of China and the Soviet Union. The United States and the newly formed United Nations (UN) bolstered democratic South Korea, led by Syngman Rhee. The tensions between the two nations ignited when the North Korean People's Army poured into South Korea in June 1950.

The United States was not prepared for an air war in Korea. There would be no nuclear exchanges, no long-term strategic targets, and it would not be fought by the theories, practices, and technologies represented by SAC. The American military and its UN allies would fight a limited, conventional war of attrition where the advanced technology of the Free World opposed the numerical superiority of the Communist Bloc.

As American and South Korean forces retreated to the Pusan Perimeter in southeastern Korea, it was the job of military aviation to protect them. The air force's Far East Air Forces (FEAF) and the navy's Fast Carrier Task Force 77 used first-generation jets, such as the Lockheed F-80 Shooting Star and the Grumman F9F Panther, and propeller-driven fighter-bombers, including the Douglas AD Skyraider and the war-surplus Vought F4U Corsair and North American F-51 Mustang, to fly a wide range of combat missions, including bomber escort, close air support, and reconnaissance. Despite the lack of adequate ground support training, these aircraft were extremely effective in attacks on tanks, troops, convoys, or fixed installations carrying various combinations of bombs, rockets, and napalm, a weapon using jellied gasoline, mounted on wing racks. By August, UN forces had destroyed all North Korean combat aircraft. MacArthur's brilliant amphibious invasion at Inchon on the western shore of Korea near Seoul in September 1950 with air cover from six aircraft carriers facilitated a UN breakout from the Pusan Perimeter and a rout of North Korean forces.

The breakout at Pusan and the success of the Inchon landings encouraged the UN commander, Gen. Douglas MacArthur, and U.S. president Harry S. Truman to change the overall aim of the war, which was to limit the spread of communism, to reunifying all of Korea as a democratic nation. UN ground forces raced to the Yalu River, the traditional boundary between Korea and China, as American and British fighters achieved air superiority and FEAF B-29s bombed the bridges spanning the river to prevent communist troops and supplies from entering or leaving the country. It appeared that a UN victory was imminent.

A new fighter jet in North Korean markings, the Mikoyan-Gurevich MiG-15, rose to oppose the B-29s and threatened to destroy UN air superiority in November 1950. The appearance of MiG-15 was a shock to UN forces because Soviet aeronautical capability had been dramatically underestimated. In the immediate post–World War II period, the Mikoyan-Gurevich (MiG) Design Bureau, founded by Artem Mikoyan (1905–1970) and Mikhail Gurevich (1892–1976) in 1939, designed a new jet fighter for the USSR. The MiG-15 featured a German-inspired swept wing and a Klimov VK-1 centrifugal-flow turbojet that was based directly upon a Rolls-Royce design that the British sold to the Soviets before diplomatic relations deteriorated in the late 1940s. With a maximum speed of 670 miles per hour and three powerful cannon, the MiG-15 was a state-of-the-art fighter interceptor.

Operating from bases located in communist China, which was off-limits to UN forces, the MiG-15s forced the B-29s to bomb at night. The introduction of the MiG-15 also signaled a new era in air combat. Air force F-80 Shooting Star pilot Lt. Russell J. Brown shot down a MiG-15 in the first all-jet dogfight in history in November 1950. Despite Brown's victory, the MiG was more than a match for the F-80 and the navy's F9F Panther. It was not until December 1950 that an aircraft capable of opposing the MiG-15 arrived in Korea. The North American F-86 Sabre, America's first swept-wing fighter, was equally as fast as the MiG with its General Electric J47 axial-flow turbojet engine and as deadly with six machine guns. Operating with the 4th Fighter Wing, Sabre pilots, many of them World War II aces, opposed North Korean, Chinese, and Soviet aviators in a battle for air superiority in the northwestern portion of the country over what became widely known as MiG Alley. By war's end, the Americans amassed a kill ratio of ten to one that destroyed almost 800 MiGs at the cost of 80 Sabres.

As Sabres and MiGs dueled above MiG Alley, UN ground forces raced toward the Yalu. Communist China committed large numbers of ground troops that moved into Korea, encircled UN forces at the Chosin Reservoir, and effectively stopped the advance. As American and British military units fought their way out, carrier aircraft supported them on the battlefield. One of those pilots was Ensign Jesse L. Brown (1926–1950), America's first black naval aviator. During World War II, African Americans fought in segregated units, but during the Korean War they served alongside their fellow white soldiers, sailors, and airmen. President Truman desegregated American military forces in 1948 through Executive Order 9981, which opened up flying opportunities not only in the new air force, but also in the navy and marines. Stationed aboard the USS *Leyte*, Brown was killed in action providing close air support in his F4U Corsair fighter-bomber to marines fighting along the Chosin Reservoir in December 1950.

The previously dynamic Korean battlefront degenerated into a stalemate near the 38th Parallel, and protracted peace negotiations at Panmunjon began in 1951. While UN ground forces worked to stop any further Communist advance, UN airpower worked to affect the peace talks with campaigns targeting North Korea's ability to fight the war. Task Force 77, in cooperation with air force and marine air units, implemented Operation Strangle to stop the movement of Communist troops and supplies between the Yalu and the 38th Parallel. James A. Michener captured the drama and futility of air attacks on heavily defended bridges and other installations in his novel (1953) and film (1954) *The Bridges at Toko-Ri*. The air force's "air pressure" campaign during the summer of 1952 targeted hydroelectric plants and expanded to include major towns and cities. By the time of the June 1953 armistice, UN airpower virtually destroyed every major North Korean city and town and obliterated the agricultural country's irrigation systems, but Korea was still divided at the 38th Parallel.

As the war raged in Korea, a new high-altitude, intercontinental jet bomber intended for SAC, the Boeing B-52 Stratofortress, took to the air in April 1952. The B-52 was larger than the B-47 with a 185-foot wingspan, faster with a cruising speed of 530 miles per hour, and capable of carrying up to 60,000 pounds of bombs or missiles for 6,000 miles. The

Stratofortress entered service with SAC in 1955. In January 1957, three B-52s accomplished the first nonstop around-the-world flight by jet-powered aircraft in less than two days.

Eight Pratt & Whitney J57 turbojets powered the B-52. As one of the most important postwar turbojet engines, the J57 powered most of America's frontline aircraft during the early Cold War. Possessing an international reputation for excellence in the design of reciprocating piston engines, Pratt & Whitney, under the direction of chief engineer Leonard S. "Luke" Hobbs (1896–1977), had struggled to develop a turbojet engine immediately after World War II, but the J57 was a clear winner. The new axial-flow engine offered increased fuel economy and twice the amount of thrust at over 10,000 pounds due to its pioneering two-spool design, which added an extra compressor and turbine into the engine for more power. The installation of a J57 enabled the first of the all-new "century series" fighters, the North American F-100 Super Sabre, to become the first airplane capable of supersonic speeds in level flight in 1953. The revolutionary performance of the J57 earned Hobbs the 1952 Collier Trophy.

Another aircraft powered by the J57 was the Lockheed U-2 photographic reconnaissance airplane, or "spy plane." The Skunk Works' combination of a lightweight airframe resembling a sport glider with the J57 resulted in high-altitude performance reaching 70,000 feet. During the 1950s, airplanes provided the air force and the Central Intelligence Agency (CIA) the important ability to photograph military and industrial locations within the Soviet Union, the People's Republic of China, and other communist countries. Unfortunately, Soviet fighters and missiles could reach the subsonic U-2 high in the atmosphere. The downing of CIA pilot Francis Gary Powers (1929–1977) over the Soviet Union in May 1960 led to increased tensions during the Cold War after Premier Nikita Khrushchev exposed the attempted cover-up of the flight by the United States. Later, in October 1962, another U-2 was shot down over Cuba after the discovery of Soviet nuclear missiles on the island nation. The resultant Cuban Missile Crisis brought the United States and the Soviet Union to the brink of nuclear war.

The Skunk Works provided a replacement for the U-2, the Lockheed SR-71 Blackbird, which could fly higher and faster than any Soviet fighter or missile by cruising at Mach 3 at altitudes above 85,000 feet, or 16 miles, near the upper edge of the earth's atmosphere. In that unforgiving environment, a SR-71 crew, consisting of a pilot and a reconnaissance systems officer, wore pressure suits similar to those worn by astronauts. The unarmed reconnaissance airplane featured a sleek delta wing that spanned 55 feet, a sinister-looking fuselage 100 feet long, and a height of 18 feet.

Designed by the Skunk Works in the 1950s, the Lockheed SR-71 Blackbird is still the world's fastest jet-powered airplane. National Air and Space Museum, Smithsonian Institution (SI 93-11885).

The Blackbird received its name from the special paint that covered its outside surfaces. Along with the titanium alloy structure, the paint allowed the skin of the airplane to withstand high-speed aerodynamic heating, caused by the friction of the air passing over the surface, near 600 degrees Fahrenheit. The black paint also absorbed radar signals and served as camouflaged for the aircraft as it flew in the dark sky at high altitudes.

Two Pratt & Whitney J58 turbojet engines, each capable of generating 30,000 pounds of thrust, powered the Blackbirds. One of the most advanced turbojets in the world, the J58 underwent a six-year development effort before entering production in 1964. The J58 relied upon innovative features based on advanced thermodynamic design to generate 30,000 pounds of thrust at high Mach speeds and operate efficiently at high temperatures. The combination of advanced aerodynamic and propulsion design made the Blackbird the fastest piloted aircraft in history with air-breathing engines. In March 1990, a Blackbird flew from Los Angeles to Washington, D.C., in 1 hour and 4 minutes at an average speed of 2,124 miles per hour.

The navy launched the carrier USS *Forrestal* in December 1954, which initiated a series of innovations that facilitated the use of jet aircraft. Emulating advances pioneered by the British, *Forrestal* featured an angled flight deck that allowed dedicated landings and takeoffs without the danger of hitting parked aircraft. Steam catapults enabled the launching of the heavy jets and a new landing system consisting of lights and mirrors was more responsive for the faster jets in all types of weather. The introduction of the nuclear-powered USS *Enterprise* in 1961 enabled long cruises, produced steam for the catapults, and removed the distracting smoke from conventional ship engines that interfered with flight operations.

A new generation of naval aircraft set sail with the new aircraft carriers. The J57-powered Chance-Vought F-8 Crusader joined the fleet in 1955 and was the first carrier-based aircraft to exceed 1,000 miles per hour and the last to be designed with guns as the primary armament. The Crusader's variable-incidence wing pivoted up to increase pilot visibility and incorporated leading and trailing edge slats and flaps for increased maneuverability for landing. The subsonic Grumman A-6 Intruder served as the navy's all-weather long-range strike aircraft beginning in 1963. The pilot and navigator-bombardier sat side-by-side in a large cockpit and used advanced avionics to follow rugged terrain and to locate enemy targets at night. Once the Intruder located the enemy, it had 15,000 pounds of ordnance to use against the target.

The United States possessed some of the most advanced aircraft in the world when it became involved in the Vietnam War (1965–1973) to limit the spread of communism in Southeast Asia. The people of Indochina defeated French colonial forces at Dien Bien Phu in 1954. The Geneva Accords of the same year established present-day Cambodia, Laos, and two Vietnams, a communist state in the north and a democratic state in the south. Ho Chi Minh, the nationalist revolutionary leader of the north, had been fighting for Vietnamese independence since 1919 and aimed to unite the two countries. The United States intended to counter the spread of communism by supporting South Vietnam with limited military advisors and equipment beginning in 1961. American involvement escalated with the arrival of jet aircraft after the 1964 Gulf of Tonkin Incident and the large-scale deployment of ground troops began the following year.

Fully committed to protecting the independence of South Vietnam, the United States was confident that its advanced military technology would easily overwhelm the North Vietnamese. During Operation Rolling Thunder, air force and navy jet fighters and bombers attacked strategic military, industrial, and transportation targets in and above the demilitarized zone (DMZ) separating the two countries beginning in March 1965. Rolling Thunder was a limited campaign intended to influence a negotiated peace with the North Vietnamese and not to provoke Communist Chinese or Soviet military intervention. President Lyndon B. Johnson stopped the campaign periodically in May and December 1965 and permanently in October 1968 to get the North Vietnamese to participate in peace negotiations and to yield to domestic and international pressure.

The air force, navy, and marines used supersonic fighter-bombers, each capable of carrying more ordnance than a World War II-era multi-engine Boeing B-17 Flying Fortress bomber, to attack the North Vietnamese. Originally designed to deliver a nuclear weapon at high speed, the Republic F-105 Thunderchief, nicknamed the "Thud," flew the majority of air strikes in the early stages of the war. The premiere fighter-bomber was the two-seat McDonnell F-4 Phantom II. Originally ordered by the navy to serve as a Mach 2 fleet defense interceptor in 1961, the F-4 offered an unprecedented flexibility for use by the air force and marines as well. The Phantom served as a reconnaissance aircraft, bomber, and air superiority fighter and used a variety of weapons, including conventional and precision-guided munitions, radar-guided Sparrow and heat-seeking Sidewinder air-to-air missiles, and rocket and heavy cannon pods. The sight of a Phantom in the air was unmistakable due to the twin black contrails from its General Electric J79 turbojets and the distinctive upward-turned wingtips and drooping rear horizontal stabilizers.

The North Vietnamese reacted to the American bombing campaigns with a coordinated network of Soviet technology. Defensive perimeters around the capital, Hanoi, and other important locations featured batteries of surface-to-air missiles (SAMs) and antiaircraft artillery. North Vietnamese aviators took to the air in subsonic MiG-17 and supersonic MiG-21 interceptors, the latter equipped with heat-seeking Atoll air-to-air missiles, to oppose American fighter pilots. There were aces on both sides. The air force's Steve Ritchie (1942–) and the navy's Randall Cunningham (1941–) both scored five victories in F-4s. Nguyen Van Coc was the highest-scoring North Vietnamese pilot with nine victories.

As the politically charged strategic air war raged north of the DMZ, the tactical air war in South Vietnam was primarily a helicopter war, but airplanes played a major role in protecting American and South Vietnamese soldiers and marines. They faced Viet Cong guerrillas and conventional North Vietnamese divisions in a highly mobile jungle war that lacked clear lines of battle. Air force, navy, and marine units operated propeller-driven and jet fighter-bombers—equipped with rockets, bombs, and napalm—as flying artillery. It appeared that in 1967, American airpower provided the firepower and mobility needed to win the war. The January 1968 Tet offensive was a stunning military defeat for the North Vietnamese, but it was a political disaster for Johnson due to its coverage on the evening news back in the United States.

In reaction to a major North Vietnamese ground offensive in 1972, President Richard M. Nixon renewed the politically motivated strategic bombing campaign through Operation Linebacker (May–October 1972) and Linebacker II (December 1972). The Linebacker campaigns depended on SAC B-52s and fighter-bombers equipped with precision-guided missiles to strike a broader range of targets, primarily Hanoi and the major port, Haiphong, which ruined the North Vietnamese economy in exchange

A SAC Boeing B-52 Stratofortress releases its bombs over North Vietnam. National Air and Space Museum, Smithsonian Institution (SI 90-15092).

for heavy combat losses. In January 1973, peace talks resumed in Paris and led to a cease-fire agreement. As soon as U.S. forces withdrew from South Vietnam in 1973, North Vietnamese and Viet Cong units overran the South, which led to the reunification of the country in March 1975. The Vietnam War was a turning point in American history. For American airpower, the war weakened the belief in superior technology and doctrine as the key to victory.

AIRPOWER, 1975–PRESENT

The American military learned a valuable lesson from the Vietnam conflict. The fighter-bomber, used in conjunction with precision-guided munitions, electronic countermeasures to ward off SAMs, and in-flight refueling, would become the dominant aerial weapons system for both strategic and tactical missions. Overall, the new aerial weapons systems were to emphasize flexibility in a variety of global missions that included, but were not limited to, the use of nuclear weapons.

A transition occurred in air force leadership. The "bomber generals" that originated with Billy Mitchell, Hap Arnold, and Curtis LeMay gave way to a distinct "fighter mafia." These command pilots shaped the introduction of new all-weather air superiority fighter-bombers. The near-Mach 3 McDonnell-Douglas F-15 Eagle entered service in 1974. The highly maneuverable Eagle was the first airplane capable of accelerating in a vertical climb because the combined thrust of its two Pratt & Whitney F100 jet engines exceeded the weight of the airplane. Fully loaded for combat, the F-15 carried a multibarrel cannon, eight missiles, and 15,000 pounds of ordnance. The performance of the Eagle was not cheap at $10 million each.

Using a single F100 engine, the Mach 2 General Dynamics F-16 Fighting Falcon was a lightweight and cheap fighter that contained interchangeable parts with the F-15. Entering service in 1979, the F-16 was the first production military aircraft with a fly-by-wire control system. Conventional aircraft relied upon a system of wires and linkages for control. The fly-by-wire system used electrical circuits and servo-actuators to control the airplane in flight. To further accentuate aerial dogfighting, an F-16 pilot sat in a reclined position with the control stick to the side with a panoramic canopy that provided unprecedented visibility in flight.

The navy introduced the gargantuan nuclear-powered *Nimitz*-class aircraft carriers in 1972 as the key component of the modern carrier battle group. Virtually at a moment's notice, the floating airfields and their protective surface and aerial screen could arrive in a crisis area and dominate

Multi-role capability and in-flight refueling made fighter-bombers, such as the F-16 Fighting Falcon, global weapons. General Dynamics via National Air and Space Museum, Smithsonian Institution (NASM 00049067).

the military situation. The presence of American carrier battle groups ranging the earth's oceans was not only symbolic of American power, but also revealed the flexibility provided by carrier operations.

The navy had its own multi-role aircraft in the 1970s, the twin engine, two-seat Grumman F-14 Tomcat. The Tomcat fulfilled the missions of air superiority and fleet defense, antishipping strikes, and ground attack. The Tomcat had a variable-sweep wing that provided flexibility for carrier operations. For takeoff and landing, the wings would be straight for low-speed maneuverability. Once in the air and pursuing a target at supersonic speeds, the pilot could sweep the wings back for high-speed efficiency. To achieve the variety of tasks related to navigation, target acquisition, electronic counter measures, and weapons employment, the F-14 had a crew of two consisting of a dedicated pilot and a specialized radar intercept officer, or RIO.

Echoing one of the major lessons of Vietnam, these new fighter aircraft incorporated technologies that facilitated aerial dogfighting. They carried air-to-air missiles and powerful multibarrel cannon for close combat. The aircraft incorporated the new head-up display, called HUD for short, that projected in front of the pilot all of the critical information required for aerial combat and flight management. The navy and the air force ensured that their pilots knew how to use these weapons by establishing the now-famous fighter weapons schools, Top Gun at Miramar Naval Air Station near San Diego and Red Flag at Nellis Air Force Base near Las Vegas. Students received classroom training in modern theories of energy management for better aerial maneuvering, pioneered by officers such as Maj. John Boyd (1927–1997) of the air force and the history of aerial combat dating back to Oswald Boelcke. Outside the classroom, instructor pilots flew light and nimble aircraft similar to Soviet fighters against students in their F-14s, F-15s, and F-16s. One of the most popular films of the 1980s, *Top Gun* (1986), captured the drama and action of low-level jet combat at Miramar.

The intercontinental strategic bomber did not become obsolete with the ascendance of the fighter mafia. The much anticipated replacement for the B-52, the Rockwell International B-1 Lancer, flew first in 1974. The B-1 featured a variable-sweep wing like the F-14, could carry twice the payload of the B-52, and generated a smaller radar profile due to its smooth blended-wing body design, but it became mired in a funding controversy during the administration of President Jimmy Carter (1924–) and did not enter service until 1982. Its primary mission was to fly low and fast under Soviet radar to reach its target.

The modification of the aging fleet of Boeing B-52s with turbofan engines (see Chapter 7 for a discussion of the turbofan engine) generated near-supersonic speeds and extended their range to more than 10,000 miles. Since World War I, strategic bombers flew directly over their targets to destroy them, which put the attacking crews at considerable risk. The B-52 became the delivery platform for a new type of weapon, the air-launched cruise missile (ALCM), which did not require the bomber to expose itself to direct enemy fire. A B-52 would launch the ALCM far from an enemy target. Once in the air, the ALCM used terrain-following radar and a light and compact turbofan engine to reach its target virtually undetected. With the ALCM and continuous upgrades to the avionics systems, the B-52 is expected to be in air force service until 2040.

New aircraft for tactical operations on the battlefield emerged. In the spirit of its World War II predecessor, the purpose-built Fairchild A-10 Thunderbolt II carried 16,000 pounds of bombs, rockets, napalm and automatic multibarrel cannon with shells containing uranium that were capable of ripping through a heavily armored Soviet tank. To ensure that the A-10 made it back to its base, Fairchild mounted the two General Electric TF34 turbofan engines above the fuselage and enclosed the cockpit in a tub of titanium armor.

The marines required a versatile aircraft that could operate in forward areas from unimproved airfields. The British military introduced the Hawker Siddeley Harrier vertical or short takeoff-and-landing (V/STOL) aircraft, or "jump jet," in 1969. The Harrier switched from hovering to forward flight via the directional nozzles of its Rolls-Royce Pegasus turbofan. The Royal Navy, operating from special aircraft carriers with special launching ramps, used the Sidewinder-armed Harriers to great effect during the 1982 Falklands War in the South Atlantic. Through an international partnership with British Aerospace and Rolls-Royce, McDonnell Douglas produced the AV-8B Harrier II, which entered service with the marines in 1985.

Instead of flying above or under enemy radar to avoid detection and possible interception by enemy fighters and antiaircraft, stealth technology made an airplane invisible. The all-black Lockheed F-117 Nighthawk was the first aircraft to incorporate stealth technology when it first flew in 1981. The F-117 featured sharp edges and angled surfaces that diverted radar waves away from their source, but were not aerodynamically streamline. The irregular shapes made the aircraft unstable and fly-by-wire technology provided the minute control movements that a human pilot could not compensate for while flying. Radar deflection took precedence over good aerodynamic characteristics. The development of the F-117 was top secret until its combat debut in Panama during Operation Just Cause in December 1989. Within its internal bomb bays, the Nighthawk carried laser-guided precision-guided munitions that it used for an unprecedented first-strike capability.

The Northrop Grumman B-2 Spirit stealth flying wing reflected another approach to aircraft design. First of all, it was a pure flying wing that lacked a fuselage and empennage. Company founder Jack Northrop worked to develop a practical flying wing throughout his illustrious career, which culminated in the unsuccessful YB-49 jet bomber of 1947. In the B-2, fly-by-wire flight management systems solved the stability and control problems Northrop faced. Regarding stealth technology, Northrop engineers used radar-absorbent coatings and infinitely curved surfaces that reflected radar signals from their source. The resultant blended body configuration of the B-2 differed greatly from the jagged edges of the F-117. The leading edge technology came with a price. At the time of its first flight in July 1989, the B-2 was the most expensive airplane in history at $2.2 billion each.

The new airplanes and weapons came together as a distinct American airpower system during conflicts in Iraq (1991), Kosovo (1999), Afghanistan (2001–2002), and again in Iraq (2003–). The iron curtain fell in the wake of the collapse of the Soviet Union in the early 1990s, which left the United States the remaining superpower. Without the threat of a global nuclear assault, American airpower, working with Allied nations, conducted joint operations in a world community. During Operation Desert Storm, network television news agencies presented vividly the forty-three-day air campaign against Iraq with striking footage of "smart bombs" destroying enemy bunkers with deadly precision. United Nations and North Atlantic Treaty Organization (NATO) air and naval forces forced the capitulation of Serbian president Slobodan Milošević of Serbia, effectively stopping ethnic cleansing in Kosovo during Operation Allied Force. In response to the terrorist attacks on the United States on September 11, 2001, American airpower pursued Taliban fighters in the mountains of Afghanistan and delivered a "shock and awe" campaign in preparation for a full-scale invasion of Iraq.

A new type of airplane, the remotely piloted uninhabited aerial vehicle, emerged during these campaigns, which could provide information about the enemy without subjecting human crews to danger. UAVs originated during World War II. They served as primitive cruise missiles and gunnery targets. The loss of Francis Gary Powers and his Lockheed U-2 over the Soviet Union in May 1960 stimulated the development of reconnaissance UAVs in the United States. The air force used the first UAV, the swept-wing Ryan AQM-34 Firebee, for missions over North Vietnam and China during the Vietnam War. The Israel military developed UAVs after the surprise Egyptian attack that began the Six-Day War in June 1967 and introduced the Scout UAV during the 1982 invasion of Lebanon. Seeing the potential value of UAVs, the U.S. Navy selected an improved version of the Scout, called Pioneer, for use as an observation platform with the fleet in 1984. The marines and the army quickly adopted their own versions.

A Pioneer RQ-2A UAV is a pusher airplane. At the end of its central fuselage is a 26-horsepower, two-cycle engine that can keep the UAV in the air for more than 5 hours. Two thin booms extending rearward connected the fuselage to the tail with its twin vertical stabilizers. Composite materials, primarily carbon-fiber, fiberglass, and Kevlar, as well as aluminum and balsa wood made up the structure of the Pioneer, which reduced its ability to be detected by radar. Pioneers carried up to 100 pounds of surveillance equipment in the fuselage, including high resolution, black-and-white television, infrared, or color cameras, or chemical, electronic, and radiological monitoring sensors. A lightweight fixed, tricycle landing gear supported the Pioneer on the ground and gave it the ability to taxi on a runway. All of that came in a small package. Overall, the Pioneer had a wingspan of 16 feet, sat 4 feet above the ground, and weighed less than 300 pounds.

The Pioneer traveled disassembled in four pieces to a combat zone in a specially built container. Two individuals could assemble it and prepare it for launching from a special platform aboard a naval warship or straight from an airfield runway. A small detachable rocket engine added extra takeoff thrust. Once in the air, two ground operators could manually control the Pioneer or set an autopilot capable of guiding it along a predetermined flight path. If needed, soldiers on the battlefield could take over control of the Pioneer. When the mission was over, the Pioneer would fly directly into a large net set up on a warship or use its built-in tail hook to catch an arresting wire set up across a runway.

Operation Desert Storm, the removal of Iraqi forces from Kuwait by a coalition of U.S.-led forces in 1991, witnessed the first major operational use of the Pioneer UAV. The marine and army UAVs aided artillery units on the battlefield. The navy Pioneers, flying from the battleships *Iowa* and *Wisconsin*, assisted in directing naval gunfire. During the bombardment of Faylaka Island in the Persian Gulf, a Pioneer flew low over the island to assess the damage made from the giant guns of the USS *Missouri*. In anticipation of another attack, Iraqi troops on the island waved white surrender flags toward the Pioneer. It was the first time that human beings had surrendered to a machine in combat.

UAVs increased in sophistication with the introduction of the Predator. General Atomics Aeronautical Systems of San Diego, California, received a Department of Defense contract for ten Predator aircraft in January 1994. The new design was a pusher monoplane with an upside-down V tail. With its turbo-supercharged 105-horsepower Rotax 914 engine, the Predator could cruise at almost 90 miles per hour for over 15 hours. A turret mounted on the bottom forward area of the fuselage housed live video and still picture cameras and radar imaging equipment capable of collecting information in all types of weather during the day or night. Ground controllers relied upon direct or satellite data links to control the Predator and to share the gathered information with battlefield commanders on a real-time basis. They went into operation from July 1995 to March 1996 in Bosnia. Going beyond the original reconnaissance role, the air force began arming Predators with laser-guided air-to-ground missiles in February 2001 for offensive operations. Later in April, the jet-powered Northrop Grumman RQ-4A Global Hawk proved the capability of high-altitude, long-endurance unmanned aerial reconnaissance when it flew 7,500 miles from the United States to Australia.

The high cost of developing state-of-the-art military aircraft contributed to the need for international cooperation between military, governmental, and industrial organizations. The supersonic Lockheed Martin F-35 Joint Strike Fighter was intended to replace a variety of military aircraft (including the F-16 Fighting Falcon, A-10 Thunderbolt II, and AV-8B Harrier) for nine separate nations. The new fighter came in three versions. The air force would operate the conventional-takeoff-and-landing version, which flew first in October 2000. The navy would use a version designed for carrier operations. The Marine Corps and the British Royal Air Force and Royal Navy intended to use the short takeoff and vertical landing (STOVL) version. In July 2001, the prototype STOVL performed a "hat trick" at Edwards Air Force Base when it successfully made a short takeoff, reached the speed of sound, and then slowed down to hover and make a vertical landing. The innovative F135 propulsion system, a cooperative development between Pratt & Whitney and Rolls-Royce, featured a dedicated lift fan and an articulated rear exhaust that enabled the F-35 to hover.

7

The Commercial Airplane, 1945–Present

◆

In October 1958, just over thirty years from when Charles A. Lindbergh made his solo transatlantic flight, a Pan American Airways Boeing 707 airliner initiated jetliner service from New York to Paris. The sleek, streamline 200-passenger airplane with its swept-back wings and four turbojet engines hanging in nacelles underneath was the result of a long-term trend that began with the DC-3 of the first Aeronautical Revolution and the new high-speed innovations that came from the second. While not the first jet airliner, the 707 ushered in a new era in mass transportation where more and more people traveled increasingly by air. The manufacture, sale, and operation of jet airliners was a big business and for the majority of the second half of the twentieth century, it was an American venture.

THE AGE OF THE PROPELLER AIRLINER, 1945–1960

World War II gave a significant boost to the world airline industry, but specifically for American air carriers. During the war, the army air forces ran a regularly scheduled aerial supply route from India to China beginning in early 1942 after the Japanese captured the Burma Road. The route, over the seemingly endless and uncharted Himalayan Mountains, kept China in the war. The crews of the Douglas C-47 Skytrain (the military version of the successful DC-3) and Curtiss C-46 Commando twin-engine transports called the experience "flying the Hump" as they flew their precious cargo over and in between mountain peaks. They faced extreme conditions ranging from heavy storms and winds to the prospect of crash-landing in the icy mountains at 17,000 feet or in the dense, tropical jungle below. By 1945, the crews were delivering 71,000 tons of supplies a month. Many of the pilots who flew the Hump as well as bomber pilots and other operators of multiengine aircraft would pursue careers in the postwar civilian airline industry. World War II also ensured that there would be a large supply of war surplus aircraft and paved runways, which opened new markets for America's leading air carriers, Pan American Airways, Trans World Airlines (TWA—formerly Transcontinental and Western Air), and United Airlines.

Continual innovation by the aviation industry would ensure that the propeller-driven airliner would be a refined technology, with tricycle landing gear, four engines, pressurized cabins, and capable of high-altitude flight at transcontinental and transoceanic ranges. Additionally, the passengers would travel in the utmost comfort and luxury. Three important four-engine airliners emerged during this period. Boeing merged the wing, tail, and landing gear of the highly successful B-29 Superfortress with the bulbous fuselage of its new military cargo airplane and aerial tanker, the C-97, to create the Stratocruiser in 1947. The aircraft had a main passenger deck with a bar and lounge on a lower level and could carry upwards of 100 passengers. Douglas Aircraft continued its successful DC series of modern airliners. The DC-4 served primarily as a World War II transport, designated the C-54, but the postwar DC-6 (1947) led to the impressive DC-7 of 1953. The final version, the DC-7C, was the first airliner capable of flying the Atlantic Ocean nonstop upon its introduction in 1956. Developed in secret at the behest of TWA's major shareholder, millionaire aviator Howard R. Hughes (1905–1976), the Lockheed Constellation was another aircraft capable of both nonstop transatlantic and transcontinental flight. The Connie, known for its distinctive triple tail, could carry more than fifty passengers. The enlarged and expanded Super Constellation, or Super Connie, could transport 100 passengers at distances up to 4,000 miles.

The engines, propellers, and fuels that powered the Boeing, Douglas, and Lockheed airliners represented the highest developments in propulsion technology. The Wright Turbo-Compound Cyclone, an improved version of the B-29's R-3350 radial engine with almost double the horsepower, powered the DC-7 and the Super Constellation. Wright engineers increased the power of the engine through the addition of three turbines

Airliners, such as this TWA Lockheed Constellation, represented the highest development of propeller-driven piston-engined aircraft. National Air and Space Museum, Smithsonian Institution (NASM A-3609).

that converted the heat from the exhaust into energy that a gearing system transmitted back to the engine crankshaft. The largest and most complex piston engine ever produced by Pratt & Whitney was the Wasp Major, which powered the Stratocruiser and generated more than 4,000 horsepower. It was popularly known as the "corncob" due to the spiral arrangement of its twenty-eight cylinders into four rows of seven cylinders for increased cooling. As an indicator of the times and the growing importance of the turbojet, the Wasp Major was also the last aircraft piston engine produced by Pratt & Whitney.

Variable-pitch propellers manufactured by Hamilton Standard and Curtiss Propeller of Caldwell, New Jersey, offered a new feature called reversible pitch that allowed heavily loaded airliners extra braking power while landing. High-octane fuels, made affordable by the high expenditures made by the U.S. government during World War II, enabled efficient high-speed cruising at 20,000 feet. The American airlines chose to continue with propeller-driven, piston–engine aircraft because they believed the jet engine was too expensive to operate, and primarily still in a developmental phase that could only be justified for military operations.

These factors contributed to a travel revolution that was well under way by the late 1950s in the United States and Europe. The airlines and the aircraft manufacturers staged aggressive marketing campaigns in leading magazines such as *Holiday*, *Newsweek*, and the *Saturday Evening Post* to popularize air travel for both business and pleasure. The airplane replaced trains and buses as the leading mode of transportation between America's cities and supplanted the steamship for travel across the Atlantic to Europe.

THE JET AIRLINER

The replacement of the piston engine with a gas turbine to drive a variable-pitch propeller resulted in the turbine propeller, or turboprop, engine. A turboprop was efficient in terms of thrust and fuel economy and it generated less noise than both piston and turbojet engines. First developed in Great Britain, turboprop airliners began commercial operations in the early 1950s and were considered to be an economical alternative to turbojet airliners for short- to medium-distance routes. The Vickers Viscount, powered by four Rolls-Royce Dart turboprops, first flew in 1948 and was the first turboprop airliner to enter passenger service. The Viscounts were popular with both European and American airlines because, compared to piston-engine aircraft, they were vibration-free, smooth in flight, and very comfortable at high-altitudes. American carrier Capital Airlines operated Viscounts in the United States. The Dutch-designed high-wing Fokker F-27 Friendship monoplane, featuring twin Darts, debuted in 1955.

The first American turboprop airliner was the Lockheed Electra II, which entered commercial service with American Airlines in December 1958. Four Allison 501 turboprop engines powered the Electra. A design flaw allowed the engine nacelles to oscillate, flutter, and vibrate, which caused catastrophic wing failure before the problem was solved.

Despite the prominence of the propeller-driven, piston-engine airliner and the possibilities of the turboprop in the 1950s, the promise of increased speeds, distances, and power and the resultant increase in revenues that the gas turbine engine offered to commercial airline operations spurred the development of a jet airliner. The first was the British de Havilland Comet. It was a sleek, streamline design with slightly swept wings and a pressurized cabin. It entered passenger service in May 1952 with British Overseas Airways Corporation (BOAC) on the London-to-Johannesburg, South Africa, route. Passengers noticed that flying in the Comet at altitudes up to

The first jet airliner: the de Havilland Comet. British Airways via National Air and Space Museum, Smithsonian Institution (NASM 00035925).

40,000 feet at speeds approaching 480 miles per hour ensured a smoother flight than did the propeller-driven airliner. Jet engines were quieter, vibration-free, and they reduced flight times to distant cities like Johannesburg and Singapore by half.

The Comet was not without its design flaws. The four Rolls-Royce Ghost centrifugal-flow turbojet engines buried in wings near the fuselage made it difficult for maintenance personnel to access it for service. In 1954, a fatal design flaw in the shape of the cabin windows and their susceptibility to fatigue after frequent pressurization and depressurization for each flight led to several catastrophic accidents. BOAC took the Comets out of service and de Havilland reintroduced an improved Comet 4 in 1958. By then, however, Great Britain had lost its lead in the international jet airliner market to the United States.

Boeing Aircraft, primarily a military aircraft manufacturer, gambled $15 million of its own money to enter the new jet airliner market under the guidance of company president William M. Allen (1900–1985). Boeing engineers took the innovations pioneered on the B-47—the highly streamlined fuselage, swept wing, and engine pods—and added a low-mounted wing configuration and tricycle landing gear to better accommodate passengers and cargo, which remained the primary arrangement for jet airliners through the twentieth and twenty-first centuries. The brown, red, and gray prototype, the Model 367–80, better known as the Dash 80, took to the air in July 1954. Boeing demonstrated its potential to customers when the famous test pilot Alvin M. "Tex" Johnston (1914–1998) performed an unauthorized barrel roll in front of thousands of excited spectators at an August 1955 boating event on Lake Washington in Seattle. Boeing quickly sold almost 1,500 of the production airliners, designated the 707, and the military transport/aerial tanker version called the KC-135. The original version of the 707 could fly 100 miles per hour faster than the Comet.

Pan American Airways, led by pioneer airline executive Juan Trippe (1899–1981), introduced the 707 on its New York-to-London route in September 1958 and ushered in the commercial jet age. The British and the Soviets had designed, built, and operated commercial jet airliners earlier, but it would be the 707 and the new all-jet Douglas DC-8, capable of carrying 150 to 180 passengers, that American airlines would use to revolutionize world air travel. These airliners initiated an almost fifty-year American domination of the world air transportation industry. American manufacturers, primarily Boeing, Douglas, General Electric, and Pratt & Whitney, produced almost 90 percent of the world's jet airliners and the engines to power them.

Four Pratt & Whitney JT3 turbojets powered the 707 and the DC-8. They were the commercial version of the highly successful J57 military

Airlines, such as Braniff International Airways, used the Boeing Model 707 to enter the jet age and expand international air travel. National Air and Space Museum, Smithsonian Institution (SI 84-3712).

engine capable of producing 10,000 pounds of thrust each. Turbojets consumed large amounts of fuel, which affected airline revenues. The addition of a large, enclosed multiblade fan to a turbojet harnessed the efficiency of the propeller while developing the high thrust of the turbojet. The unprecedented 60 to 80 percent leap in efficiency increased thrust, improved fuel economy, and reduced noise. The introduction of the new "turbofan" was crucial to the success of low-cost, long-distance air transportation.

The first practical American turbofan engine, GE's CJ805-23 aft-fan, was ready for flight in 1959. It resulted from considerable pioneering research and financial expense, primarily in a new analytical engineering procedure called computational fluid dynamics. Unfortunately, the airplane it was destined to power, the Convair 990 airliner, was not ready. Though initially resistant to the idea of the turbofan, Pratt & Whitney beat GE to the market by installing a fan on the front of a JT3 and calling it the JT3D turbofan, which generated 18,000 pounds of thrust. The airlines quickly adopted the engine for their 707 and the DC-8 aircraft in 1960. Pratt & Whitney dominated the early market for commercial turbofan engines in the early 1960s with more than 8,000 JT3D engines produced. In the 1990s, Pratt & Whitney, GE, and Rolls-Royce in Great Britain introduced turbofan engines capable of generating 100,000 pounds of thrust.

The availability of more powerful turbofan engines enabled the simple fuselage extension, or "stretching," of existing airliner designs. For the airlines, the more passengers they carried, the more revenue they generated, which could then be used to buy more airliners. Douglas was the first to stretch the DC-8 in 1965, which increased passenger capacity from 189 to 259.

Boeing and Douglas introduced new jetliners for short and medium routes. The four major American domestic airlines, American, Eastern, TWA, and United, introduced the 727 within the space of four months in 1964. Rather than having the engines housed in nacelles under the wings, the 727 was a trijet where Boeing engineers placed three engines on the rear fuselage. The resultant "clean wing," first used on the French Sud-Aviation Caravelle (1955), was aerodynamically efficient and facilitated the inclusion of a sophisticated system of high-lift devices, triple-slotted trailing edge flaps and leading-edge slats, which enhanced the 727's operational versatility. The 727 proved to be a popular with the airlines as it was the first to exceed 1,000 airframes produced. The equally successful Douglas DC-9, featuring twin engines mounted on the rear fuselage, entered service in 1965 and remained in production until 1983. Boeing countered with the 737 in 1968. The six-abreast passenger seating lowered per-mile costs, the engines suspended from wing pylons eased maintenance, and the high-lift system made it possible to operate from small, unimproved airfields, which attracted many international buyers. By the 1990s, the 737 exceeded all other airliners for orders.

As specific designs filled important niches within the international commercial aviation industry, Boeing engineers learned that they could no longer base new designs on the basic 707 fuselage as they had done with both the 727 and the 737. Encouraged by Juan Trippe, William Allen made another gamble like he made with the 707 and ordered the development of the first "wide-body" airliner, the 747, from design data used in a failed military transport. With a wingspan of 195 feet, a length of 231 feet, and a weight of 735,000 pounds, the new "jumbo jet" could carry nearly 500 passengers at speeds approaching 600 miles per hour at a range of 6,000 miles. The 747 entered service on Pan American's New York-to-London route in January 1970. It is still in production in the early twenty-first century and flying on long-distance routes all around the world. Through longevity and its sheer size, the 747 helped make the world become a "smaller" place. The most famous 747 is *Air Force One*, the aerial transport for the president of the United States.

The 1960s and 1970s were turbulent times for the American aviation industry. Douglas overextended itself financially with both military and commercial contracts and found that the only way to survive was to merge with the military aircraft and spacecraft manufacturer McDonnell in 1967. The new McDonnell Douglas Corporation aimed to compete in the commercial airliner market with its new DC-10 wide-body airliner. In 1968, Boeing was an aerospace giant with $3.3 billion in sales and 142,000 employees. A significant downturn in the national economy combined with cancelled military, space, and the failed aircraft contracts led to a downsizing of the workforce to 56,000 employees in 1971. The scarcity of jobs led to the erection of a roadside billboard that proclaimed, "Will the last person leaving Seattle turn out the lights?" The economy improved and Boeing diversified into nonaeronautical areas, including rail and sea transport and computers, but the core business of selling commercial airliners continued.

An international competitor to American manufacturers emerged in the 1970s. Airbus Industrie, a multinational association of British, French, German, Italian, and Spanish companies centered in Toulouse, France, introduced the A300 in 1972 for intercity European travel. Airbus introduced new innovations normally found on military aircraft on its commercial airliners. The twin-engine A320 airliner, which entered service with Air France in February 1987, featured advanced metal alloys and synthetic composites in the structure. In the cockpit, Airbus replaced conventional instruments with electronic flight information systems that used computerized visual monitors and television cathode ray tubes. The multitude of screens and displays led to its characterization as a "glass cockpit." To fly the A320, the pilots used a fly-by-wire system, complete with the side-stick controllers as found on the General Dynamic F-16 Fighting Falcon fighter-bomber. With aircraft like the A320, and the larger A330 series that could compete with the Boeing 747 and the 757 and 767 twin-jets, Airbus effectively challenged American dominance in the international commercial airliner market. The United States possessed 91 percent of the commercial aircraft market in 1969. That percentage had dropped to 67 percent in 1993.

Boeing pioneered a revolutionary new design process with the design of its wide-body twin-turbofan 777 airliner in 1991. Traditional aircraft design involved the generation of thousands of paper drawings and blueprints and full-size wood mock-up airplanes that could instantly become obsolete if a correction or modification needed to be made. Boeing engineers used high-speed digital computers and three-dimensional graphics programs to facilitate the efficient design and manufacture of the new airliner from the drawing board to the production floor. With all drawings stored in a computer, company engineers coordinated the production of subassembly components with contractors located around the world through a sophisticated digital network. Additionally, any correction to a minute design flaw could be made in an instant. When the twin-jet took to the air in 1995, it was the first "paperless airplane."

In the late twentieth century, the two primary commercial jetliner manufacturers—Boeing and Airbus—offered two choices for the airlines. In 1997, McDonnell-Douglas merged with Boeing after losing several important contracts and being surpassed by Airbus in airliner production. Airbus offered the behemoth 550 passenger A380, the world's largest airliner when it took to the air, in April 2005. The trend toward even larger aircraft reflected the desire to ease congestion of the airways and airports by carrying more people in fewer, but larger, airliners. Boeing's 787 Dreamliner, scheduled to enter service after 2008, can carry only 200 to 300 passengers, but in an aerodynamically sophisticated and environment-friendly design where half the entire structure of the airplane is composed of advanced composite materials.

For many in the world aviation community, primarily the United States, the Soviet Union, Great Britain, and France, the next logical step in commercial aviation was the development of a supersonic transport (SST) to get people to their destinations faster than ever before. The American SST program began in reaction to European intentions to construct a supersonic airliner in the 1950s. With the support of President John F. Kennedy and the Federal Aviation Administration (FAA), the U.S. government initiated a design competition between the leading aircraft manufacturers. The government chose Boeing's Model 2707 design in December 1966 and GE received the contract for the engine. Almost immediately, a national debate began that centered on the cost of development, predicted to total $5 billion, and the environmental impact. Many groups objected to the prospect of experiencing frequent sonic booms while the exhaust pollution expelled from the 2707's four turbojets would contribute to the deterioration of the ozone layer. The Senate cancelled funding for the program by a vote of 49 to 48 in March 1971.

The Soviet Union and the British and the French working together designed, manufactured, and operated SSTs in commercial service. Both designs featured a pointed nose, slender fuselage, delta wing, and afterburning jet engines, which reflected the considerable body of design knowledge that existed for supersonic military aircraft and were firsts for commercial airliners. To assist in taxiing operations, the entire nose of these new SSTs could be pointed down so that the pilots could see the runway. In the Soviet Union, the Tupolev Design Bureau's Tu-144 was the first to fly in December 1968, but a spectacular crash in front of thousands of spectators at the 1973 Paris Air Show and the high cost of operation ensured that none of the aircraft would enter long-term operational service.

The Anglo-French Concorde was the only SST to enter sustained commercial service. Great Britain and France agreed to share the overwhelming technical and financial obstacles through a formal agreement in November 1962. The Concorde first took to the air in March 1969 and made its first Mach 1 flight the following October and its first Mach 2 flight in November 1970. The world's airlines lined up to place orders for the new airliner in 1973, but the world oil embargo went into effect and stimulated an international energy crisis. The resultant high cost in aviation jet fuel ensured that the Concorde would never be an economically viable airplane. The British and French governments exerted substantial force on their respective national airlines, British Airways and Air France, to order seven of the SSTs to keep the project alive. The Concorde went into service with both in January 1976. Twenty-seven years of high-speed and high-luxury flying, with an individual seat costing as much as $10,000, continued until service ended in October 2003.

COMMERCIAL AVIATION IN TRANSITION

Women and African Americans found a place in commercial aviation in the 1970s. Traditionally, women served as flight attendants, or "stewardesses." United Air Lines hired the first female flight attendants in 1930. They served as aerial nurturers who domesticated the flight experience for predominantly male passengers. By the 1950s, serving as an airline flight attendant was a glamorous activity that was an intermediate step before marriage. With the introduction of unionization, training, and increased professionalization in the 1970s, women flight attendants pursued long-term careers in aviation. Helen Richey (1909–1947) became the first female commercial airline pilot when she joined Central Airlines in 1934. The all-male pilot's union made the experience unbearable for her and she resigned the following year. It would not be until 1973 when Emily Howell Warner joined Frontier Airlines that another female airline pilot would take to the air. Before she was through, Warner also became the first woman to earn her captain's wings while accumulating 21,000 flight hours in the cockpit. After a twenty-year career in the air force, former Tuskegee Airman Robert "Bob" Ashby joined Howell at Frontier Airlines in 1973 and retired as a captain in 1986.

The Airline Deregulation Act of 1978 led to a restructuring of the American commercial aviation system. Before 1978, the government's Civil Aeronautics Board regulated fares, routes, and schedules that allowed airlines to operate point-to-point service that connected major cities with direct flights. Deregulation created a hub-and-spoke system that on paper resembled a wagon wheel. Passengers started their journey at a regional airport and traveled down a "spoke" to the "hub," or a major airport. Once there, they would join other passengers traveling to the same city and proceed on to their final destinations. The hub-and-spoke system ensured that airlines could compete with each other in an efficient manner. The needs and comfort of travelers was secondary. Cramped airports, flight delays due to bad weather, and time wasted waiting for circuitous connecting flights characterized what some critics called the "hub-and-choke" system. As more and more people traveled by air—665 million passengers traveled the nation's airlines in 2000 alone—the once glamorous experience became less desirable. Deregulation did ensure lower air fares, which were a benefit for travelers.

Deregulation facilitated the growth of a new generation of commuter airlines operating twin-turboprop aircraft along the "spokes" of the new system in the 1980s. The large airlines abandoned these routes and these small companies emerged to meet the demand. In the 1970s and before deregulation, the world oil crisis stimulated the use of propeller-driven aircraft, due to their high efficiency and lower operating costs for the rapidly expanding short-haul commuter market. Two aircraft designed and manufactured in Canada and Brazil became instantly recognizable at America's airports. The high-wing de Havilland Canada Dash 8 appeared in 1984 and initially carried forty passengers. Brazil's leading aircraft manufacturer, Embraer, introduced the low-wing thirty-passenger EMB-120 Brasilia in 1985. Pratt & Whitney's Canadian subsidiary manufactured the powerful and fuel-efficient PW100 series of turboprops for these new aircraft, which were capable of generating 2,000 to 5,000 horsepower.

From a modest 6,000 passengers in 1926, the number of Americans traveling by air grew to 41 million in 1955, and 275 million in 1978. In 1979, there were twenty-nine national and international airlines and they carried 317 million passengers, which was 63 percent of the U.S. population. While revenues and the number of passengers increased, a sluggish national economy and a 1981 strike by the Professional Air Traffic Controllers Organization (PATCO), who sought shorter hours and higher salaries, contributed to a slump in revenues. The PATCO strike highlighted the importance of air traffic control and the demand for improving the all-important infrastructure of aviation. The FAA and IBM developed En Route Stage A, an automated system of radar and computers that tracked airliners as they traveled between airports. Advances in electronics such as airborne radar and the microwave landing system facilitated the introduction of improved navigation and landing systems.

The expansion and modernization of America's airports added to the increased sophistication of the commercial aviation system. Municipal authorities had a choice. They could improve existing facilities dating back to the 1950s, such as Chicago's O'Hare, or construct an entirely new facility like the Denver International Airport, which opened in 1995. NASA contributed a major improvement to runway design by developing a grooved surface that channeled water away to prevent airliners from hydroplaning.

In the wake of deregulation, the 1980s was an unstable time for American air carriers. The pioneering airline executives, such as the lively Robert F. Six (1907–1986) who had shaped Continental Airlines from a small regional airline into a major international carrier, were leaving the industry. Financier Frank Lorenzo (1940–) acquired Continental in 1981 and Eastern Airlines in 1986. His adept use of bankruptcy laws allowed him to recast his new acquisitions in more favorable economic terms and generate unprecedented profits, but at the expense of workers' welfare. Driven out of the airline business, Lorenzo retired with a sizable fortune and Eastern dissolved in 1991 after being severely weakened by the high fuel costs in the wake of Operation Desert Storm.

Pan American was the traditional "chosen instrument" for American international air travel since its beginnings in the 1920s. When other airlines, primarily United, expanded to international routes, the ailing airline was unable to compete. The American economic recession, and the overextension of funding regarding the purchase of the 747 fleet, and a failed attempt to expand into domestic routes hastened further problems and Pan American was forced to sell its major routes in 1990 and went bankrupt in December 1991. By 1992, three airlines, the Big Three consisting of American, United, and Delta, controlled more than 50 percent of the commercial aviation market.

Airlines became increasingly international in character after deregulation. Foreign airlines such as KLM, Scandinavian Airlines System, and British Airways invested heavily in ailing American air carriers Northwest, Continental, and US Air. This involvement led to the creation of the alliance system where partner airlines representing different regions of the globe worked together to create an integrated air travel network that benefited all parties. The airlines did not have to expand and passengers could travel to more locations around the world. American Airlines and British Airways joined together in the Oneworld Alliance in 1999 that included member airlines from Australia, Canada, Chile, China, Finland, Ireland, and Spain. In contrast to the growing trend of international integration, Herb Kelleher (1931–) and Southwest Airlines stayed independent and domestic in focus while offering low fares, no extra services, and operating only one airplane, the Boeing 737.

Congested airports, weather delays, and smoking passengers were a minor annoyance in comparison to the upsetting trend of air terrorism, which dated back to the early days of air transportation. Aerial hijacking increased in the 1960s in reaction to the tensions of the Cold War when groups commandeered commercial airliners to either leave or enter communist countries such as Cuba. Turbulent events in the Middle East, and American involvement in the region, led to an escalation of hijackings and violence in the 1980s. Lebanese Shiite Muslims hijacked TWA Flight 847 flying from Athens to Rome in June 1985. Of the 145 passengers aboard, 104 were Americans and most of them were coming home from a visit to the Holy Land. The hijackers diverted the airliner to Beirut, Lebanon, and began a two-week ordeal that resulted in the murder of one American sailor. The U.S. air raid on Libya in 1986 and the accidental destruction of an Iranian jet airliner by a navy cruiser that killed 290 people during the summer of 1988 did not alleviate the situation. In Frankfurt,

Germany, a Middle Eastern terrorist group planted a bomb aboard Pan American Flight 103 in December 1988. As the Boeing 747 left its connection in London headed toward New York, it exploded over Lockerbie, Scotland, killing all 270 aboard. These events led to an increase in government travel advisories and airport security forces, which included the use of electronic metal detectors, luggage inspection, and x-ray machines to deter further hijackings.

The trend of aerial hijackings culminated in the terrorist attacks against the United States on September 11, 2001. A highly coordinated group of Middle Eastern terrorists hijacked four American Airlines and United Airlines jetliners and used them as fuel-laden aerial bombs. They destroyed the twin towers of the World Trade Center in New York City and damaged one section of the Pentagon in Washington, D.C. Due to the efforts of the passengers, who had been made aware of the earlier attacks via cell phone, the fourth airliner, United Flight 93, crashed in the Pennsylvania countryside near Pittsburgh. The attacks left approximately 3,000 Americans dead, sent the airline industry reeling, intensified airport security procedures, and initiated a global war against terror.

8

General Aviation, 1920–Present

◆

On any given summer afternoon for the past sixty years, passersby could see a Piper Cub—a small, yellow, two-seat high-wing monoplane powered by a four-cylinder engine and a wood propeller—take to the air from a small grass airstrip or local airport across the United States. The reason for the flight was clear: for the pure enjoyment of flying by the pilot and the passenger. Flying for recreation is one of the many activities that make up general aviation, which involves a variety of aircraft performing a myriad of airborne activities. Essentially, any type of flying other than scheduled commercial airlines and military aviation, including aerial demonstration and sport, agricultural dusting, cargo transport, business travel, firefighting, flight training, and utility operations, can be considered general aviation.

GENERAL AVIATION THROUGH WORLD WAR II

World War I created a new generation of fliers in the United States and many of them wanted to continue flying in their private lives. Rejecting the danger and migratory nature of barnstorming, these early pilots settled down at local airports and became fixed-base operators (FBOs), which became the foundation for American general aviation. FBOs performed many functions, including aircraft sales and repair, flight instruction, scenic

An airplane for everyday flying: the Piper J-3 Cub. National Air and Space Museum, Smithsonian Institution (NASM A-47056-A).

tours, hunting expeditions, high-speed delivery, and passenger service. A private individual could purchase a war–surplus Jenny, rent a hangar space at the FBO, fly for fun, and socialize with local aviation enthusiasts. Thousands more Americans aspired to take to the air after Charles Lindbergh's 1927 solo transatlantic flight. By the end of the decade, there were more than 15,000 "private pilots" and many more would-be aviators waiting for flight training. Flying clubs, including ones formed at prestigious Ivy League schools such as Harvard University, joined the FBOs as centers for everyday flying.

A new generation of light airplanes, or "lightplanes," made for private flying soon emerged. The Monocoupe series of high-wing monoplanes was one of the first. Donald A. Luscombe (1895–1965), a young advertising executive from Davenport, Iowa, regularly flew his open-cockpit Curtiss Jenny on business and personal trips in the mid-1920s. He wanted an airplane with an enclosed cockpit that would allow him to wear a suit while flying rather than a leather helmet, goggles, a flight jacket, and coveralls. Unable to find what he wanted, Luscombe designed his own airplane and with funding from his fellow members in the local flying club, he started his own company in October 1926. Luscombe's first employee was Clayton Folkerts (1897–1965), a young self-taught engineer, who turned his employer's idea into a flying airplane.

Luscombe called his airplane the "Monocoupe," which stood for the combination of a monoplane wing with the concept of a two-seat sporty automobile. The first Monocoupe featured an enclosed cabin with two seats placed side by side, a welded steel tubular fuselage and tail, a wood wing, and a cotton fabric covering. Overall, the airplane was small with a wingspan of approximately 30 feet and light at nearly 800 pounds minus the pilot, passenger, and fuel. It took to the air in April 1927 and became the first light cabin monoplane to be approved for operation by the U.S. government in January 1928. Monocoupes made up approximately 90 percent of the American light airplane market in the late 1920s.

Other new manufacturers of light aircraft emerged primarily in the Midwest around Wichita, Kansas. Walter H. Beech (1891–1950), Clyde Cessna (1879–1954), and Lloyd Stearman (1898–1975) founded the Travel Air Manufacturing Company in January 1925 and then went on to start their own successful companies. WACO (short for Weaver Aircraft Company and pronounced "Wah-Co") of Troy, Ohio, produced biplanes that became synonymous with the glamour of private and corporate flying in the 1920s and 1930s. From 1931 to 1936, these companies produced a total of 5,000 airplanes, but suffered from heavy financial losses due to the Great Depression.

In many ways, mainstream America reluctantly accepted the airplane as an instrument of business, recreation, and travel. The desire of women to pursue careers in aviation merged with contemporary attitudes that they were not as able as men in all spheres of life to create an effective sales technique: "If a woman can fly, then anyone can." Light aircraft manufacturers hired women to sell their products to an enthusiastic, but timid, public. Louise Thaden (1905–1979) proved to be an effective aircraft salesperson. She won the National Women's Air Derby from Santa Monica to Cleveland in 1929, and with copilot Blanche Noyes (1900–1981) beat out many male pilots to become the first women to win the long-distance Bendix Trophy in 1936. Thaden and Noyes's airplane was a beautiful blue and white Beech Staggerwing, a luxurious personal biplane with the lower wing positioned farther ahead than the top wing, retractable landing gear, and capable of speeds of more than 200 miles per hour. Husband and wife team Walter and Olive Ann Beech (1903–1993) benefited from Thaden's success to sell their aircraft to an expanding international market. Thaden's memoirs, *High, Wide, and Frightened* (1938), told a story of adventurous flying from the perspective of a female pioneer.

Despite the image of women being inferior to men to sell airplanes, female aviators were enthusiastic about aviation. They formed their own group, the Ninety-Nines, in December 1929, which was a combination women's rights organization, business association, and social society. Consisting of only licensed pilots, the Ninety-Nines established credibility for women in aviation who wanted to assert their rights while still appearing feminine within contemporary cultural boundaries.

William J. Powell (1897–1942) saw aviation as a source of upward mobility for his fellow African Americans during an era that had little to offer minorities. He founded a flying school and aero club in honor of the first African American pilot, Bessie Coleman (1893–1926), in Los Angeles in 1929 and helped organize the first all-black air show in the United States in 1931. Powell successfully recruited celebrities in the African American community, such as the famous boxer Joe Louis, to support his efforts. His semiautobiographical book, *Black Wings* (1934), and his journal, *Craftsmen of Black Wings* (1936), urged African Americans to embark upon aviation careers ranging from aircraft mechanic to airline tycoon. Powell's ultimate goal was to create an African American air transportation system free from racial discrimination. His ideas never became reality, but they were a strong indication of how the airplane excited and moved people to improve their lives and places in society.

Powell was part of a larger African American aviation community in the United States. In Chicago, John C. Robinson organized the Challenger Air Pilots Association in 1931 and constructed the first airfield built for and by African Americans in nearby Robbins, Illinois. Cornelius R. Coffey (1903–1994), a 1931 graduate of the first all-black mechanics class at the Curtiss-Wright Aeronautical School, established the Coffey School of Aeronautics in Chicago in 1934. J. Herman Banning (1899–1933) and Thomas C. Allen completed the first transcontinental flight by black aviators in 1932. A year later, C. Alfred Anderson (1907–1996) and Dr. Albert E. Forsythe (1897–1986) flew from Atlantic City to Los Angeles and back during the first round-trip transcontinental flight by black aviators. They made a goodwill tour of South America in 1934 in an airplane named *Booker T. Washington* in honor of the African American civil rights pioneer.

The financial costs of flying lessons and aircraft ownership, however, were too high for many Americans regardless of gender or race. In Cincinnati, the Aeronautical Corporation of America, or Aeronca, introduced the single-seat C-2 monoplane in 1929 for $1,500 ($16,207 in modern currency). Civilian engineers working at the army's research and development facilities near Dayton, Ohio, designed both the C-2's airframe and its two-cylinder 30-horsepower opposed engine in their spare time. Pilots called the C-2 the "flying bathtub" for its distinctive shape and somewhat limited performance. The introduction of the C-2 initiated the creation of a new industry producing somewhat affordable aircraft for private flying.

Another way to get flying was for a mechanically inclined individual to purchase an airplane kit, consisting of drawings and some components, and build it at home. The Heath Airplane Company of Chicago offered the Super Parasol monoplane in 1929 for $199 ($2,150 in modern currency). Builders had to find their own engine, but Heath offered a small 100-pound, four-cylinder, air-cooled, 30-horsepower Henderson in-line motorcycle engine adapted for use in the Parasol. Both the C-2 and Parasol offered minimal performance, but they were lightweight and relatively inexpensive. For many of these "homebuilders," the first (and unfortunately for some, the last) time they ever took to the air was in a "homebuilt," or "kitplane," that they constructed and then taught themselves to fly. Toward the end of the 1930s, government legislation limiting private construction and flying of airplanes effectively quelled the homebuilding movement.

In the early 1930s, the Great Depression gripped America in a severe economic downturn that hurt all levels of society. In the spirit of President Franklin D. Roosevelt's pathbreaking social legislation called the New Deal, the director of the Commerce Department's Aeronautics Branch, Eugene L. Vidal (1895–1969), endeavored to promote private flying and to encourage the development of safe and affordable aircraft for personal use. Considerable funding provided for the improvement of municipal airports and the building of smaller private airfields across the country. Vidal sponsored a 1934 contest for aircraft manufacturers to develop an all-metal, two-seat monoplane that would cost $700, or almost $10,000 in today's currency. The price was about the same for an automobile, which was affordable by many Americans. Poor relations between the government and the aviation industry ended the contest and the possibility of an airplane for everyday flying at a price average Americans could afford.

Vidal persevered to encourage designers and manufacturers to develop an airplane that was easy to fly, safe, and capable of production in large numbers. One of these aircraft was the Ercoupe designed by Fred E. Weick. After his groundbreaking work on the NACA cowling, he turned his energies toward improving personal and working aircraft that were affordable, simple to fly, and safe. His W-1A monoplane reflected those goals. It had an integrated control system that combined aileron and rudder control to ease the process of learning to fly by removing one extra set of controls, the rudder pedals. The elevator had a built-in limited range of motion to prevent spins. The W-1A's most noteworthy feature was a new type of landing gear with a steerable nose wheel at the front of the airplane and two fixed wheels located behind the center of gravity. The arrangement made the airplane easier and safer to handle on the ground. Additionally, the airplane was already in a flight attitude, which meant it would get into the air that much faster. The aviation industry quickly adopted what Weick called his "tricycle" landing gear, which quickly became the predominant form for the majority of modern aircraft.

Weick incorporated those innovations into his Ercoupe, manufactured by the Engineering Research Corporation (ERCO) of College Park, Maryland. The little airplane was a two-seat, all-metal, low-wing cantilever monoplane that advertising heralded as the "world's safest plane" when it was introduced in 1940. ERCO produced 6,000 Ercoupes, and many are still flying in the early twenty-first century.

At the height of the Great Depression, oil tycoon William T. Piper (1881–1970), a major shareholder in the struggling Taylor Brothers Aircraft Company of Bradford, Pennsylvania, bought the company in 1931. The small company produced a small two-seat high-wing monoplane for $4,000, but Piper wanted the company to produce a cheaper and more economical version called the Cub for $1,500. Taylor produced more than 300 E-2 Cubs beginning in 1931. Piper still wanted to improve the basic design and hired Walter C. Jamouneau (1912–1988) in 1935. The engineer added an enclosed cabin, joined the wing to the fuselage, rounded the wingtips, and curved the empennage of the new Taylor J-2 (J stood for *J*amouneau). After a factory fire, the company moved to Lock Haven, Pennsylvania, and changed its name to Piper Aircraft Corporation in 1937. A year later, Piper introduced the J-3 Cub with its distinctive bright yellow paint, black trim, and bear cub logo for $1,300 ($16,732 in modern currency). With a wingspan of 35 feet and a length of 22 feet 4 inches, the Cub was a true lightplane weighing in at only 680 pounds. Piper had manufactured more than 14,000 Cubs when production ended in 1947. The little J-3 Cub became one of the most recognizable aircraft designs in aviation history. Many are still flying today with new generations of pilots.

A new series of aircraft engines powered the little Cubs. Continental of Detroit, Lycoming of Williamsport, Pennsylvania, and Franklin of Syracuse, New York, manufactured four-cylinder, air-cooled engines in the 40- to 65-horsepower range. The arrangement of the cylinders horizontally across from each other led them to be called opposed engines. The small, lightweight, cheap, and powerful opposed engine made private flying practical and affordable.

The creation of the Civilian Pilot Training Program (CPTP) in 1939 increased the demand for J-3s. Emulating European state-sponsored flying programs, the CPTP provided subsidized flight training through universities, colleges, and local airports. Initially, the CPTP aimed to create a new generation of young Americans who would see flying as an everyday activity and purchase light aircraft. The establishment of a large reserve of military pilots ready for additional training became the primary focus after American entry into World War II. The scope of the program included women and African Americans, which reflected the societal constraints of the time. Since women were not allowed to participate in combat operations, would-be female aviators were turned away and most of the women with previous flight experience served as instructors during the war. A completely self-sufficient segregated program existed for African American men. The CPTP trained 435,165 people to fly, which significantly accelerated the growth of private pilots in the United States. Seventy-five percent of all trainees took their initial flight instruction in Cubs.

The American military used general aviation aircraft for a variety of purposes during the war. The principal use was the training of pilots and aircrew. Aviation cadets flew the Stearman PT-13D Kaydet biplane and the Fairchild PT-19 Cornell and Ryan PT-22 Recruit monoplanes during their first phase of flight training before moving on to larger and more powerful aircraft. The twin-engine Beech AT-series provided would-be pilots experience with multiengine aircraft and navigators and bombardiers with the opportunity to practice their specialties. More than 40,000 bombardiers learned how to operate the Norden bombsight sitting in the Plexiglas nose of the AT-11 Kansan. The production of these aircraft could not interfere with production of high-performance combat aircraft, so manufacturers used nonstrategic materials such as plywood, which led cadets to call them Bamboo Bombers. Army ground forces used "grasshoppers," Piper Cubs, and similar aircraft from Aeronca, and Taylorcraft, for artillery-spotting, observation, and liaison on the front lines.

The U.S. government grounded private flying for the duration of World War II. Male and female private pilots volunteered themselves and their airplanes for the Civil Air Patrol (CAP). Male CAP pilots patrolled America's east and gulf coasts for Nazi submarines, a dangerous exercise that claimed the lives of twenty-six pilots. The CAP also performed airport management, forest-fire patrol, humanitarian and liaison flights, and search-and-rescue missions. Cartoonist Zack Mosley (1906–1994) brought the story of the CAP to mainstream America through his popular nationally syndicated comic strip, *Smilin' Jack* (1933–1973).

GENERAL AVIATION IN THE POSTWAR ERA

The aeronautical community believed that there would be a postwar general aviation boom in the United States. They attributed that to the thousands of military pilots returning to civilian life and thousands more interested in learning to fly and able to pay for it with the G.I. Bill. It appeared that the decade-old dream of "an airplane in every garage" was close to fruition. In addition to their family, home, car, and the gray flannel suit, veterans would commute to work by airplane and spend their leisure time flying in aviation country clubs. Membership in the Aircraft Owner's and Pilots Association (AOPA) began an exponential growth that would grow to include half of all private pilots in the United States. AOPA worked to portray a positive image of general aviation especially through the columns of its official publication, *Popular Aviation*. The boom was short-lived. After a record sale of more than 33,000 aircraft in 1946, orders fell dramatically and many manufacturers suffered major financial losses or went out of business altogether.

Beech, Cessna, and Piper emerged from the postwar "bust" as the primary American general aviation manufacturers. Beech introduced the Bonanza high-performance low-wing monoplane in March 1947. With a six-cylinder opposed engine and retractable tricycle landing gear, the Bonanza could carry five passengers and their luggage with ease. The Bonanza's distinctive "V," or "butterfly," tail aimed to provide the same level of control as a conventional three-surface empennage. In a modified Bonanza called *Waikiki Beech*, William Odom flew 5,273 miles, from Hawaii across the Pacific and the United States to New Jersey, in little over thirty-six hours nonstop in March 1949. The Bonanza is still in production with more than 17,000 V- and straight-tail versions produced over a fifty-year period.

Cessna introduced the four-seat, all-metal, single-engine Model 170 high-wing monoplane in 1948, which became a popular choice for private pilots. Geraldine "Jerrie" Mock (1925–) became the first woman to pilot an aircraft around the world in her tail-dragger Cessna 180, the *Spirit of Columbus*, in 1964. The new two-seat Model 150 of 1959 featured new tricycle landing gear and served a new generation of individuals yearning to earn their private pilot's licenses.

Private pilots also chose to fly antique airplanes in the 1950s. The nostalgia of a rotary-engine-powered World War I fighter or the barnstormers' Curtiss Jenny echoed the early days of flight. For pilots seeking higher performance, they could purchase a variety of World War II–surplus military aircraft, including Curtiss P-40 Warhawk and North American P-51 Mustang fighters and Boeing B-17 Flying Fortress and North

American B-25 Mitchell bombers. The exhibition of these "warbirds" became a popular attraction at air shows. The Commemorative Air Force (CAF), founded in 1951, reenacted the great aerial battles of World War II with restored military aircraft, which included spectacular pyrotechnic displays complete with patriotic music and narration.

There was a rebirth of homebuilding aircraft in the post–World War II period. In Oregon, George Bogardus (1914–1997) organized the old groups of the 1930s into the American Airmen's Association in 1946. He lobbied the government for a permanent regulatory category for experimental aircraft and used the association's magazine, *Popular Flying*, and whirlwind automobile and airplane tours to spread his message. When the legislation became a reality in 1952, Bogardus had almost single-handedly revitalized the homebuilding movement.

In Milwaukee, Wisconsin, a group of homebuilders led by national guard pilot Paul H. Poberezny (1921–) founded the Experimental Aircraft Association (EAA) in January 1953. The EAA held its first annual convention and air show, or "fly-in," the following September with more than twenty aircraft attending. By 1970, when the fly-in had to be relocated to Oshkosh, Wisconsin, it had become a major event in aviation. The grassroots character of the EAA went beyond the homebuilding community. Its journal, *Sport Aviation*, encouraged the restoration, preservation, and flying of antique military, commercial, and civilian aircraft through competition, provided outreach to the public through its educational foundation and museum, and fostered aerial sporting activities such as aerobatics.

The existence of a large, private community of pilots willing to build and maintain their own aircraft encouraged the emergence of a new kind of aeronautical engineer. Burt Rutan (1943–) introduced a technological sophistication and creativity into homebuilt aircraft that reflected his ability to integrate new innovations into his designs in a new and exciting way. Rutan earned his aeronautical engineering degree from California Polytechnic University, San Luis Obispo, in 1965 and worked for the U.S. Air Force and in the aviation industry as a flight test engineer before starting his own company, the Rutan Aircraft Factory, at Mojave, California, in 1974.

While still in college, he became interested in using a canard, a horizontal stabilizer mounted forward of the wings, which the Wright brothers used on their airplanes. The inspiration for his first design, the all-wood VariViggen, came from the Swedish Saab J-37 Viggen jet fighter, which was the first high-performance airplane to feature canards. Rutan was a firm believer in the benefits to be had from using a canard. They decreased the occurrence of stall, the sudden loss of lift, which could lead to an overall loss of aircraft control, an often fatal event. This facilitated an ease of manufacture because all control linkages and surfaces were in close proximity to the cockpit. Finally, an airplane equipped with a canard was visually distinctive and unconventional in appearance.

Rutan introduced innovative construction techniques with new materials, called glass composites, in his next canard airplane, the futuristic VariEze ("very easy"). For many years, European glider builders cut foam to specific shapes and covered them with fiberglass and epoxy glue to construct aerodynamically smooth, light, and strong airframes. Rutan adapted the methods to his design process. It was easier, and less expensive, to use composites than it was to use wood or metal when he wanted to experiment with different canard and wing configurations. The almost 3,000 homebuilders who bought VariEze plans after they went on sale in 1976 benefited from Rutan's use of glass composites because they did not need specialized mechanical knowledge or tools to work with the new materials. Once completed, a VariEze equipped with a 100-horsepower engine was capable of carrying two adults a distance of 700 miles at a speed of 180 miles per hour. Rutan went on to design other visually striking, high-performance glass-composite kitplanes, including the larger Long-EZ.

General aviation also allowed pilots to engage in sporting activities. Aerobatics provided an opportunity for pilots to demonstrate their skill at performing spectacular and astonishingly precise aerial maneuvers. Aerial displays dated back to Lincoln Beachey in the early flight period and grew in popularity in the 1920s and 1930s. While individual skill varied from pilot to pilot, there was one airplane that became the design of choice for aerobatic pilots for almost three decades after World War II, the Pitts Special S-1. Rather than rely on large, heavy, and expensive aircraft designed for other uses, Curtis Pitts (1916–2005) wanted a small, lightweight, purpose-built, and extremely agile aerobatic airplane. His innovative 1945 design featured short-span swept biplane wings, an opposed engine, and it could climb, dive, and roll better than any other aerobatic airplane in the sky.

Aerobatic champion Betty Skelton (1926–) purchased the second production Pitts Special in 1948 and named it *The Little Stinker*. She installed an additional bank indicator, an instrument that tells a pilot if the wings are level, upside-down in the cockpit to make it easier to conduct maneuvers during inverted flight. Skelton went on to win the 1949 and 1950 national aerobatic championships for women in the tiny red and white biplane with its signature skunk insignia. Her spectacular aerobatic displays showcased the capabilities of the Pitts Special and made it an instant favorite with the general aviation community. When Pitts offered a homebuilt version of his Special, called the S-1C, in the 1960s, home-builders flocked to purchase the plans, construct their own aircraft, and compete in aerobatic competitions in the United States and Europe. Quickly, the instantly recognizable Pitts Special became the most successful American aerobatic airplane in history.

Almost forty years after Betty Skelton's victories in *The Little Stinker*, much had changed for women and their airplanes in aerobatics. After 1972, men and women no longer competed in separate competitions in the United States, but in integrated contests that ignored gender. Patricia "Patty" Wagstaff (1951–) became the first woman to win in the new format when she won the first of three U.S. National Aerobatic Championships in September 1991 in an Extra 260 aerobatic monoplane. In the early 1980s, German aerobatic pilot Walter Extra (1954–) unsuccessfully competed in a Pitts Special against a new generation of midwing monoplane aerobatic airplanes based on the Stephens Akro. He decided to build his own aerobatic airplane embodying a combination of traditional construction—a tubular steel fuselage and wood wings covered in fabric—with the latest innovations, primarily a composite tail, ailerons, and landing gear. Extra purposefully designed the 260's wing to be unstable, which made the aircraft more maneuverable and easier to control during flight. At each wingtip was a special device that enhanced the accuracy of the pilot's maneuvers. A modified Lycoming six-cylinder opposed engine strengthened to withstand the stress and strain of violent maneuvers to the point of failure powered the 260. Overall, the little monoplane could climb straight up at 4,000 feet per minute and could complete one roll in one second.

Air racing is as old as the airplane. Beginning in the early flight era, daredevil aerial sportsmen used their frail contraptions to oppose each other in aerial contests similar to horse and equally new automobile races. After the bloodbath of World War I, air racing transformed quickly from an individual to an institutionalized sport where American and European military teams competed in the name of international prestige, interservice rivalry, and the advancement of technology. The next era, 1929–1939, reflected the aerial populism of Depression-era America with Horatio Alger-like individuals such as Roscoe Turner (1895–1970) taking readily available technologies and building air racers to achieve fame and fortune. The last eras, 1946–1949 and 1964 to the present, emerged in the spirit of the 1930s and evolved into an ultracompetitive and financially and socially exclusive version of motor sports dominated by heavily modified World War II fighter aircraft centered on the remote town of Reno, Nevada.

Besides flying for pleasure or sport, a new generation of working aircraft emerged. America was still a predominately agricultural nation in the early twentieth century. Oftentimes, plagues of insects destroyed entire crops and the local economies that grew and depended on them. Army researchers at McCook Field worked with the state of Ohio and the U.S. Department of Agriculture to pioneer the development of aerial spraying, or "crop dusting," beginning in 1921. They took an army Curtiss Jenny and added a metal hopper; the device was used to distribute powdered or liquid fertilizers and pesticides into the air, and sprayed orchards in the Ohio countryside to prevent a moth infestation. Additional tests at Tallulah, Louisiana, targeted the boll weevil, a major threat to the all-important cotton crop in the American South. The film *Fighting Insects from Airplanes* documented the work for interested viewers in local cinemas.

In the 1920s and 1930s, specialist companies, primarily Huff-Daland Dusters and Delta Aero Dusters (forerunner of Delta Airlines), sprang up and traveled across North and South America dusting crops. In the wake of World War II, dusting companies purchased war-surplus Stearman Kaydet biplanes, which were ideal due to their rugged structures and easy flying qualities, for as low as $250 ($2,500 in modern currency) and modified them with more powerful engines and hopper systems for aerial application. Fred Weick and a team of engineers at Texas A&M University designed the first true crop duster, the Ag-1 monoplane, in 1950, which led to the highly successful Piper Pawnee series. The new airplane featured a cockpit framed in steel for safety and mounted high on the fuselage for superior visibility, a low-wing configuration, and corrosion-resistant materials to offset deterioration caused by chemical fertilizers and pesticides delivered by a sprayer system.

Every year, wildfires destroyed hundreds of acres of forests and grasslands in the North American west. During the 1920s, 1930s, and 1940s, the U.S. Forest Service used airplanes for locating remote fires, dropping airborne firefighters, or "smoke jumpers," by parachute to battle them, and then to support the effort on the ground through aerial resupply. The Forest Service and the state of California worked to develop air tankers, airplanes capable of dispersing fire-retardant chemicals or water from the air to fight wildfires, in the 1950s. The majority of air tankers have been converted from surplus multiengine propeller-driven military aircraft from the World War II, Korea and Vietnam war eras. The qualities that made them excellent combat aircraft, long-range, the ability to carry heavy payloads, and durable structures, facilitated their use as air tankers. The first dedicated air tanker, the Canadair CL-215 amphibian, appeared in 1967. Able to operate from both water and land, the CL-215 skims the surface of a lake or river as openings in its hull collect 1,400 gallons of water in approximately 10 seconds. To say that flying an aerial tanker is dangerous is an understatement. To get the water or fire-retardant chemicals to a blazing forest or grassland, the air tanker must fly at altitudes in the area of 100 feet, which subjects it to turbulent and superheated wind conditions.

While crop dusters and air tankers flew dangerously low to the ground, corporate executives traveled above the clouds at high speeds in airplanes offering the same comfort and luxury they experienced at five-star hotels. The use of the airplane for business travel offered the flexibility to travel anywhere in the United States in an age of multistate operations. In the 1920s and 1930s, corporations such as Standard Oil bought the first airplanes, primarily converted commercial airplanes, for executive travel. Beech introduced one of the first aircraft designed for business travel, the eight-to nine-passenger Model 18 Twin Beech, in 1937, and it remained in continuous production until 1969. A quarter of a century later, Beechcraft introduced the low-wing, all-metal King Air series, which featured a pressurized cabin, climate control, soundproofing, and two Pratt & Whitney PT6 turboprop engines. Able to carry six to eight passengers, the King Air quickly became one of the most popular business aircraft in history because it could fly higher and faster than piston-engine aircraft while being able to operate out of small airports.

The first jet designed specifically for business aviation was the Learjet. William P. "Bill" Lear, Sr. (1902–1978), inventor of the first practical autopilot, had a successful business in converting propeller-driven aircraft into long-range executive aircraft. While living in Switzerland, Lear began work in 1959 on a business jet that would appeal to buyers with its speed and small size. His new jet resembled a modern jet fighter because the design team used the thin wing design from a failed Swiss fighter-bomber and added wingtip fuel tanks. Propulsion came from two General Electric turbojets that the company derived from a successful fighter engine series. With a wingspan of 35 feet and a length of 43 feet, the compact Learjet could carry two pilots and six passengers. The Learjet utilized advances found in other areas of aeronautics, including the area rule fuselage, wing spoilers for speed control, an adjustable horizontal stabilizer, and a pressurized cabin. Lear chose to move his company to Wichita, Kansas, the home of America's general aviation industry, and the first Learjet took to the air in the fall of 1963. The new jet could cruise at 500 miles per hour at an altitude of 45,000 feet while flying high above bad weather and over long distances.

The Learjet led to a new generation of business jets. Cessna introduced the seven-passenger Citation in 1971. The Citation had a range of 1,200 miles with a cruising speed of 350 miles per hour and could operate out of the same airports as the King Air. With two Pratt & Whitney Canada JT15D turbofans generating 4,400 pounds of thrust, the Citations were economical to operate, quiet, and, above all else, affordable. An improved model of the Citation, capable of near-Mach speeds while carrying fourteen passengers, debuted in 1993.

Individuals and companies used general aviation aircraft for additional activities over the course of the twentieth century. Reflecting their utility, these aircraft operated from lakes and rivers in the Alaskan and Canadian bush country, conducted oil surveys in remote desert regions, surveyed urban development and new territorial boundaries, and served as aerial delivery trucks. Providing a new perspective on the earth below, aircraft served as vehicles for exploration. In 1931, an American Geographical Society expedition discovered the Great Wall of Peru, a major archaeological find.

GENERAL AVIATION AND THE FUTURE OF FLIGHT

The earliest representations of flight dating back to Leonardo da Vinci envisioned the pilot providing propulsion. In 1959, the wealthy British industrialist Henry Kremer offered a monetary prize for the first sustained and controllable flight made under human power. Dr. Paul B. MacCready (1925–) and a team of dedicated engineers at AeroVironment began work toward building an airplane specifically designed for the purpose in 1976. The new airplane, called the Gossamer *Condor*, consisted entirely of aluminum tubing, a clear plastic covering called Mylar, cardboard and foam for the wing's leading edges, and stainless steel bracing wires. The pilot, champion bicyclist and hang glider aficionado Bryan Allen, generated a third of a horsepower for propulsion, which traveled to the single propeller via a bicycle crank and chain. Allen controlled the Gossamer *Condor* with a lever that controlled vertical and lateral movement and a forward horizontal stabilizer and wing-warping system that harkened back to the Wright brothers. The Gossamer *Condor* was a large airplane with a wingspan of 96 feet, length of 30 feet, and a height of 18 feet, but the total weight of the airplane, minus the 137-pound pilot, was only 70 pounds.

After a series of trial-and-error experiments, the Gossamer *Condor* traveled little over one mile as it flew a figure-eight pattern around two pylons in 6½ minutes at a speed of 11 miles per hour at Shafter airport in California in August 1977. MacCready and his team successfully won the

Paul MacCready's human powered Gossamer Condor in flight with Bryan Allen at the pedals and controls. National Aeronautics and Space Administration (NASA) via National Air and Space Museum, Smithsonian Institution (SI 84-6698).

Kremer Prize, which by then had grown to £50,000, or $95,000, for designing and building the first human-powered aircraft capable of sustained and controllable flight. MacCready went on to win a second Kremer Prize and a Collier Trophy when he developed the Gossamer *Albatross*, the first human-powered aircraft to fly across the English Channel in June 1979.

MacCready and AeroVironment moved from human power to solar power with their next designs. The Solar Challenger was the first solar-powered aircraft to fly in June 1981. The ultimate goal was to create new unpiloted vehicles that could perform flights of long duration of up to six months and at high altitudes of 50,000 to 100,000 feet for atmospheric research and even perform the same duties as orbital satellites. Power came from solar cells that converted the energy of the sun to drive electric motors that turned the propellers. The unmanned Pathfinder, essentially a flying wing with no fuselage, featured six propellers and electric motors and climbed to 80,000 feet in August 1998 and the follow-on design,

Helios, flew to almost 97,000 feet and set a world altitude record for propeller-driven aircraft.

Burt Rutan left the homebuilt market in 1982 and created a new company dedicated to full-scale commercial products, Scaled Composites. He used the techniques learned from his homebuilt designs to build a number of innovative new aircraft that overcame specific aeronautical challenges. Rutan's most well-known design was Voyager, a long-distance airplane intended to fly nonstop around the earth without refueling. Voyager was a three-boom airplane with long, thin wings that stretched out over 100 feet. The center boom, or fuselage, housed the cockpit, cabin, main fuel tank, and two engines mounted on either end. The two outward booms housed additional fuel tanks. Constructed almost entirely of composite materials, the entire airplane weighed approximately 900 pounds, but could carry more than 7,000 pounds of fuel at takeoff.

Voyager's propulsion system was unique, which highlighted its purpose-built design. The rear engine, a 110-horsepower liquid-cooled Teledyne Continental IOL-200 known for its extreme fuel efficiency,

Burt Rutan's futuristic Voyager circled the Earth in nine days. National Air and Space Museum, Smithsonian Institution (SI 2001-7671).

operated during the entire flight. The more powerful front engine, a 130-horsepower, air-cooled Teledyne Continental 0–240, ran only when Voyager needed extra power. Variable-pitch propellers manufactured by the Hartzell Propeller Company of Piqua, Ohio, maximized the power of the engines for efficient cruising.

After almost a decade of development, training, and fund-raising, Burt's brother Dick (1938–) and Jeana Yeager (1952–) took off from Edwards Air Force Base, California, and flew Voyager nonstop around the world over the course of nine days in December 1986. They covered almost 25,000 miles at an average speed of 116 miles per hour. The Voyager team received the 1986 Collier Trophy for its excellence in design, manufacturing, planning, and skill.

Rutan and Scaled Composites went on to pioneer additional innovative aircraft. The futuristic Beech Starship corporate airplane featured a canard, forward-swept wing, and a twin pusher configuration, but was a commercial failure due to its $5 million price tag. Turning his attention toward private access into space, he designed *Spaceship One* and its launch vehicle *White Knight*, which won the Ansari X-Prize of $10 million for the first private flight into the earth's atmosphere in 2004.

Conclusion: Higher, Ever Higher

The future holds many possibilities for the airplane. At NASA's Dryden Flight Research Center in California, a McDonnell-Douglas F/A-18 fighter with a computer-controlled wing that flexes, twists, and changes its shape in a manner similar to birds flew in March 2003. This new aeroelastic wing is a modern version of the wing-warping mechanism the Wrights used on their Flyer that has the potential to enhance performance for a variety of aircraft. Other new innovations include artificially intelligent fly-by-wire flight control systems that allow aircraft to stay aloft after a catastrophic structural failure, new aerodynamic designs that reduce or altogether eliminate sonic booms, and new hypersonic propulsion systems that will enable airplanes to fly into orbit.

For the three major areas of aviation—commercial, military, and general—changes appear to be on the horizon. With the exception of jet engines and swept wings, commercial airliners are visually and structurally the same as the revolutionary DC-3. The blended flying-wing configuration pioneered on the Northrop Grumman B-2 Spirit offers opportunities for gigantic aircraft capable of carrying people and cargo deep within its structure over long distances. The UAV, capable of extreme combat maneuvers that would kill a human pilot, presents a significant challenge to military manned flight. A future generation of stealth UAVs armed with precision-guided weapons and air-to-air missiles would be more than able to fight future wars similar to the late twentieth and early twenty-first century conflicts in Eastern Europe and the Middle East. The dream of "an airplane in every garage" persists with many enthusiastic engineer-entrepreneurs, who identify advanced aerodynamics, new and efficient propellers and piston engines, and state-of-the-art avionics as key to their success.

At a particularly frustrating point in the development of the first airplane in 1901, Wilbur Wright exclaimed to his brother Orville, "Not within a thousand years will man ever fly." Two years later, they succeeded in creating the world's first airplane. In the 100 years since, humankind integrated the airplane fully into modern life as a vehicle for research, travel, business, leisure, and war. This international quest for higher, faster, and farther in aeronautical technology will ensure that the airplane will continue to evolve in the same way humankind has evolved.

Glossary

Aerodynamics. A branch of fluid dynamics that deals with the movement of the airplane and the forces acting upon it in relation to airflow.

Aileron. Movable surface hinged on the outer trailing edge of a wing used for roll control.

Airfoil. The aerodynamic shape of a wing or propeller that produces lift.

Airframe. The structure of an airplane.

Amphibian. An airplane that is able to operate from both water and land.

Aspect ratio. The measure of how long and slender a wing is from tip to tip and its relationship to aerodynamic efficiency.

Avionics. A combination of *avi*ation and electro*nics*; the electronic instruments and control equipment used in airplanes.

Biplane. An airplane with two wings mounted one above the other.

Camber. The curvature of an airfoil from front to back.

Canard. A horizontal stabilizer mounted forward of the wings.

Cantilever. A structure that requires no external bracing.

Center of gravity. The point in an aircraft's structure where its weight is evenly distributed between front and rear.

Cockpit. Where a pilot operates the controls and systems of an airplane.

Composite. Structural materials that combine two or more organic or inorganic components.

Cylinder. Engine component that contains the explosion of the fuel-air mixture during combustion.

Delta wing. A triangular wing designed for high-speed flight.

Dihedral. A condition in which the wingtips are located higher than the center line of the airplane.

Drag. The aerodynamic force that reduces the forward motion of an airplane in flight.

Elevator. Movable surface hinged on the outer trailing edge of a horizontal stabilizer that provides longitudinal pitch control, or the ability to go up or down.

Empennage. The tail of an aircraft including the vertical and horizontal stabilizers and their control surfaces.

Fillet. Fairings that smooth the flow of air between the fuselage and wings.

Fixed landing gear. Consists of wheels and struts set in a permanent landing position.

Flap. Movable surface mounted on the inside trailing edge of a wing that generates high lift at low speeds and increased drag during landing.

Flutter. Oscillations that result from an aerodynamic imbalance.

Flying boat. An airplane with a hull designed for marine operation.

Flying tail. An adjustable horizontal stabilizer that can be moved up and down for better control at transonic speeds.

Flying wing. An airplane without an empennage or fuselage.

Fuel system. Combines air with fuel to create an inflammable gas for combustion in an engine.

Fuselage. The central structure that supports the lift-producing wings and tail and houses the pilots, passengers, cargo, instruments, engine, fuel, and landing gear.

High-wing monoplane. An airplane with the wing sitting on top of the fuselage.

In-line engine. An engine where the cylinders are arranged along the length of the crankshaft.

Laminar flow. Condition where air moves smoothly over the top of a wing in evenly compressed layers while producing maximum lift and minimum drag.

Leading edge. The front edge of a wing.

Low-wing monoplane. An airplane with the wing sitting at the bottom of the fuselage.

Mach. The speed of an aircraft in the air in relation to the speed of sound.

Midwing monoplane. An airplane with the wing attached to the center of a fuselage.

Monocoque. Construction method where the outer skin of the fuselage does not require internal support.

Monoplane. An airplane with one wing.

Nacelle. A streamlined structure that usually houses an engine and protrudes from a wing.

Opposed engine. An engine where the cylinders are arranged horizontally across from one another.

Piston, connecting rod, and crankshaft. Engine components that transmit the energy of combustion into rotary power.

Pitch. The angle in which an airplane travels through the air.

Propeller. Rotating wing that transmits the energy created by the engine into thrust to propel an airplane forward.

Pusher. A propeller mounted at the rear of an airplane.

Radial engine. An engine with the cylinders arranged around the crankcase.

Reversible-pitch propeller. A variable-pitch propeller capable of reversing thrust.

Roll. The banking and tilting motion of an airplane in flight.

Rotary engine. An engine where the propeller, cylinders, and crankcase rotate on a stationary crankshaft.

Rudder. Movable surface located at the rear of the vertical stabilizer that permits directional control from side to side.

Slats and slots. Devices found on the leading and trailing edge of a wing used to enhance lift.

Spin. Condition where an airplane enters into an unstable and uncontrollable oscillation.

Streamline. The process of reducing drag to generate higher cruising speeds and lower fuel consumption.

Superchargers and turbosuperchargers. Force air into the cylinders to increase combustion and improve performance at high altitudes.

Thrust. The force that moves an aircraft through the air.

Tractor. A propeller mounted at the front of an airplane.

Tricycle landing gear. Consists of a steerable nose wheel at the front of the airplane and two fixed wheels located behind the center of gravity.

Triplane. An airplane with three wings mounted one above the other.

Valve. Engine component that allows the vaporized fuel-air mixture to enter the cylinder and exhaust gases to exit.

Variable-pitch propeller. Allows the angle, or pitch, at which each propeller blade rotates through the air to vary according to different flight conditions.

Wind tunnel. A specialized research tool used to generate aerodynamic data such as lift and drag in a controlled environment.

Wingspan. The length of a wing from tip to tip.

Wing-warping. A system of lateral control where one end of an airplane wing flexes up and the other down.

Yaw. The side to side directional movement of an airplane in flight.

Selected Bibliography

Aeronautics Division of the National Air and Space Museum. *Aircraft of the Smithsonian*. 2004. http://www.nasm.si.edu/research/aero/aircraft/.

Anderson, David F., and Scott Eberhartdt. *Understanding Flight*. Washington, DC: McGraw-Hill, 2001.

Anderson, John D., Jr. *The Airplane: A History of Its Technology*. Reston, VA: AIAA, 2002.

———. *Introduction to Flight*. 5th ed. Boston: McGraw-Hill, 2005.

Bednarek, Janet R. Daly, and Michael H. Bednarek. *Dreams of Flight: General Aviation in the United States*. College Station: Texas A&M University Press, 2003.

Bilstein, Roger E. *Enterprise of Flight: The American Aviation and Aerospace Industry*. Washington, DC: Smithsonian Institution Press, 2001.

———. *Flight in America: From the Wrights to the Astronauts*. 3rd ed. Baltimore: Johns Hopkins University Press, 2001.

Biography Files. National Air and Space Museum Archives, Smithsonian Institution, Washington, DC.

Boeing Company. *Boeing: The History Beneath Our Wings*. 2005. http://www.boeing.com/history/flash.html.

Boyne, Walter J. *The Messerschmitt Me 262: Arrow to the Future*. Washington, DC: Smithsonian Institution Press, 1980.

Conner, Margaret. *Hans von Ohain: Elegance in Flight*. Reston, VA: AIAA, 2001.

Constant, Edward W. *Origins of the Turbojet Revolution.* Baltimore: Johns Hopkins University Press, 1980.

Coopersmith, Jonathan, and Roger Launius, eds. *Taking Off: A Century of Manned Flight.* Reston, VA: AIAA, 2003.

Corn, Joseph J. *The Winged Gospel: America's Romance with Aviation.* New York: Oxford University Press, 1983.

Craven, Wesley F., and James L. Cate, eds. *The Army Air Forces in World War II.* 7 vols. Chicago: University of Chicago Press, 1948–58. Reprint, Washington, DC: Office of Air Force History, 1983.

Crouch, Tom D. *Wings: A History of Aviation from Kites to the Space Age.* New York: W. W. Norton, 2003.

Curry, Andrew. "Taking Wing: A Century of Flight." *Smithsonian* 34 (December 2003). http://www.smithsonianmag.si.edu/smithsonian/issues03/dec03/pdf/flight.pdf.

Douglas, Deborah G. *American Women and Flight since 1940.* Lexington: University Press of Kentucky, 2004.

The Epic of Flight, 23 vols. Alexandria, VA: Time-Life Books, 1980–1983.

Galison, Peter, and Alex Roland, eds. *Atmospheric Flight in the Twentieth Century.* Boston: Kluwer Academic Publishers, 2000.

General Electric Transportation. *General Electric Aircraft Engines.* 2005. http://www.geae.com/engines/index.html.

Gibbs-Smith, Charles H. *Aviation: An Historical Survey from Its Origins to the End of World War II.* 3rd ed. London: Science Museum, 2003.

Golley, John. *Genesis of the Jet: Frank Whittle and the Invention of the Jet Engine.* Shrewsbury, UK: Airlife, 1996.

Gross, Charles J. *American Military Aviation.* College Station: Texas A&M University Press, 2002.

Gunston, Bill. *The Development of Jet and Turbine Aero Engines.* Somerset, UK: Patrick Stephens, 1997.

Hallion, Richard P. *Strike from the Sky: The History of Battlefield Air Attack, 1911–1945.* Washington, DC: Smithsonian Institution Press, 1989.

Hansen, James R. *The Bird Is on the Wing: Aerodynamics and the Progress of the American Airplane.* College Station: Texas A&M University Press, 2004.

Hansen, James R., Jeremy R. Kinney, J. Lawrence Lee, and David Bryan Taylor. *The Wind and Beyond: A Documentary Journey through the History of Aerodynamics in America.* Washington, DC: Government Printing Office, 2003.

Hardesty, Von. *Black Aviator: The Story of William J. Powell and a New Edition of Black Wings.* Washington, DC: Smithsonian Institution Press, 1994.

Hardesty, Von, and Dominick Pisano. *Black Wings: The American Black in Aviation.* Washington, DC: National Air and Space Museum, Smithsonian Institution, 1983.

Heppenheimer, T. A. *A Brief History of Flight: From Balloons to Mach 3 and Beyond.* New York: John Wiley & Sons, 2001.

Jakab, Peter L. *Visions of a Flying Machine: The Wright Brothers and the Process of Invention*. Shrewsbury, UK: Airlife, 1990.

Kranzberg, Melvin. "Technology and History: 'Kranzberg's Laws.'" *Technology and Culture* 27 (July 1986): 544–560.

Launius, Roger D., ed. *Innovation and the Development of Flight*. College Station: Texas A&M University Press, 1999.

Launius, Roger D., and Janet R. Daly Bednarek, eds. *Reconsidering a Century of Flight*. Chapel Hill: University of North Carolina Press, 2003.

Loftin, Lawrence K., Jr. *Quest for Performance: The Evolution of Modern Aircraft*. Washington, DC: National Aeronautics and Space Administration, 1985.

Millbrooke, Anne. *Aviation History*. Englewood, CO: Jeppesen Sanderson, 2000.

Milstein, Michael. "Hyper-X: Inside the Fastest Air-Breather on the Planet." *Air & Space Smithsonian* 20 (June–July 2005): 46–51.

National Academy of Engineering of the United States of America. *Memorial Tributes*. 10 vols. Washington, DC: National Academy Press, 1979–2002.

National Aeronautics and Space Administration. *Back to the Future: Active Aeroelastic Wing Flight Research*. 2005. http://www.nasa.gov/centers/dryden.

National Aeronautics and Space Administration History Division. *Aeronautics*. 2005. http://www.hq.nasa.gov/office/pao/History/index.html.

National Air and Space Museum. *Black Wings: African American Pioneer Aviators*. 2002. http://www.nasm.si.edu/interact/blackwings/index.html.

National Aviation Hall of Fame. *Enshrinees*. 2005. http://www.nationalaviation.org.

National Museum of the United States Air Force. *The Aircraft and History of the United States Air Force*. 2005. http://www.wpafb.af.mil/museum/.

Rolls-Royce. *1904–2004: A Century of Innovation*. 2004. http://100.rolls-royce.com/index.jsp.

Schlaifer, Robert, and Samuel D. Heron. *Development of Aircraft Engines and Aviation Fuels*. Boston: Harvard University Graduate School of Business Administration, 1950.

United States Centennial of Flight Commission. *Essays on the History of Flight*. 2003. http://www.centennialofflight.gov/.

Van Riper, A. Bowdoin. *Imagining Flight: Aviation and Popular Culture*. College Station: Texas A&M University Press, 2004.

Winter, Frank H., and F. Robert Van der Linden. *100 Years of Flight: A Chronicle of Aerospace History, 1903–2003*. Reston, VA: AIAA, 2003.

Wohl, Robert. *The Spectacle of Flight: Aviation and Western Imagination, 1920–1950*. New Haven, CT: Yale University Press, 2005.

Young, James O. *Lighting the Flame: The Turbojet Revolution Comes to America*. Edwards AFB, CA: Air Force Flight Test Center History Office, 2002.

Index

JEREMY R. KINNEY is Curator in the Aeronautics Division at the National Air and Space Museum of the Smithsonian Institution. He has served as the Centennial of Flight lecturer at the University of Maryland.